2D and 3D Graphene Nanocomposites

2D and 3D Graphene Nanocomposites

Fundamentals, Design, and Devices

edited by

Olga E. Glukhova

CRC Press
Taylor & Francis Group
Boca Raton London New York

CRC Press is an imprint of the
Taylor & Francis Group, an **informa** business

2D and 3D Graphene Nanocomposites: Fundamentals, Design, and Devices

First published 2020 by Jenny Stanford Publishing Pte. Ltd.

Published 2023 by CRC Press
Taylor & Francis Group
6000 Broken Sound Parkway NW, Suite 300
Boca Raton, FL 33487-2742

ISBN 978-981-4800-41-9 (hbk)
ISBN 978-0-429-20150-9 (ebk)

Visit the Taylor & Francis Web site at
http://www.taylorandfrancis.com

and the CRC Press Web site at
http://www.crcpress.com

British Library Cataloguing-in-Publication Data
A catalogue record for this book is available from the British Library.

Contents

Preface

The discovery of graphene by A. K. Geim and K. S. Novoselov in 2004 divided the modern history of materials science into *before* and *after*. The research boom that began after the release of the first publication on graphene has affected many strategically important high-tech industries: energy conversion and energy storage, environmentally friendly technologies, medicine, electronics, and space applications. One of the new stages in the development of graphene materials science is associated with the study of composite materials based on graphene and other carbon nanostructures, including carbon nanotubes and fullerenes. A large number of scientific papers are devoted to theoretical and experimental studies of composite materials based on 2D graphene and 1D carbon nanotubes. There are several topological types of graphene–carbon nanotube composites that vary in the method of joining nanotubes and graphene as well as their mutual orientation. 3D pillared vertically aligned carbon nanotube–graphene composites and graphene–carbon nanotube composite films with horizontally aligned nanotubes are of most interest to researchers. The reason for such increased attention to graphene–carbon nanotubes composite materials is the improvement of their properties compared with conventional carbon materials due to the synergistic effect of nanotubes and graphene. The first results of research of new carbon composite structures showed their superiority over individual nanotubes and graphene in electrical, optical, and electrochemical properties, which opens up new ways to develop promising practical applications based on these materials.

Hybrid carbon nanosystems consisting of C_{60} fullerene films deposited on a monolayer graphene have broad prospects. Reports about this material began to appear in scientific papers only a few years ago, however, it is already predicted that electronic nanodevices, including photovoltaic devices, will be developed on the basis of graphene/C_{60} hybrid systems. It is predicted that a new promising direction in electronics will be the development of sensitive graphene-based sensors of single biomolecules, in particular, the DNA molecule. An important task in this area is to study the electronic energy and electrically conductive properties of graphene/DNA nucleotide molecular complexes from the standpoint of identifying the greatest sensitivity of graphene for the efficient operation of a DNA sensor.

An interesting and promising kind of graphene material is glass-like carbon, which is called an intermediate carbon material between layered graphene and fullerenes. This material has an irregular highly porous structure and a highly developed surface. There are two topological types of glass-like carbon: graphitized type—a material from multi-layered graphene flakes forming layers, and non-graphitized type—a material consisting of randomly distributed curved layers of graphene flakes and fragments of layered fullerenes, which form numerous nanopores. The presence of a porous structure makes this material suitable for filling with atoms of various chemical elements in order to control its characteristics in a wide range of values.

Numerous 2D graphene-like materials have already been distinguished into a separate class of materials, in the structure of which either carbon atoms in the lattice sites are replaced by atoms of a different chemical type or various adsorbed atoms are present. One of the main purposes of such modifications is the opening of the band gap in graphene, the value of which can be controlled, for example, by varying the concentration of adsorbed atoms. In particular, the functionalization with hydrogen and fluorine atoms leads to the appearance of a band gap in graphene. The new graphene modifications obtained in this way—hydrogenated graphene (graphane) and fluorinated graphene (fluorographene)—are intensively investigated by many scientific groups.

The progress achieved in the research of graphene and graphene-based composite materials to date would not be possible without computer simulation methods that significantly complement the results of full-scale experiments and in some cases can replace them due to the special complexity or impossibility of performance. With the emergence of new structural varieties of graphene, the physical and mathematical apparatus is also being developed for their study.

This book covers the fundamental properties of new 2D/3D composite materials based on graphene and its modifications and discusses their potential application areas. It provides an interpretation of interesting physical effects discovered for the first time for graphene materials under consideration. It contains a description of the experimental methods for obtaining and characterizing samples of chemically modified graphene and the conceptual foundations of popular methods for computer modeling of graphene nanostructures as well as original computational techniques developed by the authors of the book chapters. The book is a result of the joint work of many highly qualified specialists in the field of experimental and theoretical research of graphene and its derivatives. The book will be useful for a wide audience, including students, graduate students, and researchers in academic institutions and also specialists in industrial engineering.

Olga E. Glukhova
Summer 2019

Chapter 1

Graphene and Graphene Nanomesh-Based Platforms for Sensing, Transparent Electrodes, and Catalysis Templates

Ahmed A. Maarouf[a] and Amal Kasry[b]

[a]*Department of Physics, Institute for Research and Medical Consultations, Imam Abdulrahman Bin Faisal University, PO Box 1982, Dammam 31441, Saudi Arabia*
[b]*Department of Basic Science, Faculty of Engineering, the British University in Egypt (BUE), El-Sherouk City, Suez Desert Road, Cairo 11837, Egypt*
amaarouf@iau.edu.sa

In this chapter, we discuss graphene-based templates for three main applications: biological and chemical sensing, photovoltaics, and catalysis. The three applications utilize the electrical, optical, and chemical properties of graphene-based structures.

In biosensing, a change in the electronic and/or optical properties of graphene is the core sensing process. Initially, some surface modification has to occur to increase the reactivity of graphene, and hence the stability, accuracy, and reproducibility of its sensing properties. This can be done in an electrostatic way, through the binding of a biological material, while retaining the electronic properties of graphene. Compared to metals, graphene has a strong

2D and 3D Graphene Nanocomposites: Fundamentals, Design, and Devices
Edited by Olga E. Glukhova

ISBN 978-981-4800-41-9 (Hardcover), 978-0-429-20150-9 (eBook)
www.jennystanford.com

fluorescence quenching effect, whose mechanism can be utilized for biological sensing.

A graphene sheet has a transparency of 98%; therefore, a doped graphene sheet can be used as a transparent electrode. Industrial grade graphene contains grains, which have low sheet resistance. The sample as a whole has an impractically high sheet resistance because of the grain boundaries. To create parallel paths for electrons in a graphene sheet, metallic nanotubes can be deposited on the graphene surface, thus creating a graphene/carbon nanotubes (CNTs) hybrid material with high transparency and industrially appealing sheet resistance.

Graphene nanomeshes (GNMs) are porous graphene structures formed by fabricating pores in a graphene sheet. Depending on the details of the pore lattice, GNMs can be semiconducting, with a fractional eV electronic gap. The gap can be utilized to form a charge sensor, in which the sensing mechanism is based on the boost of the electronic conductance of the GNM. The pore edges can be selectively passivated to form electrostatic traps for charged nanoparticles, which may dock into the passivated pores. Such a template is best suited to catalytic nanoparticles, as it prevents their agglomeration.

1.1 Introduction

Graphene, since its isolation in 2004 [25, 26], has attracted a tremendous amount of interest because of its unique chemical and physical properties [70, 101], including high electronic mobility, optical transparency, and mechanical strength. Recent advances in large-area growth and isolation of graphene have established it as a promising candidate for several technological applications [44, 52, 53], such as a platform for biosensing [97], transparent electrodes [9], supports for catalysis [63, 96], and nanoelectronics [19, 81].

1.2 Graphene in Biosensing Applications

Biosensors are considered a key medical technology, aiming at the identification of chemical or biological compounds that provide diagnostic and patient monitoring solutions for increasing the efficiency of the healthcare system. The key element of a typical biosensor is the transducer, which converts a biological interaction into measurable signals; according to the type of the transducer, the signal can be optical, electrical, mechanical, or electrochemical. The huge growth in the biosensor market drives the development of new transducers that are sensitive, fast, specific, and stable, can perform in real time, and are based on inexpensive materials. In recent decades, there has been much progress in transducer technologies; one of the most successful and efficient biosensor techniques is based on surface plasmon resonance (SPR). SPR utilizes the phenomenon that incident light of a specific wavelength can be absorbed in a thin metal layer (typically gold or silver) to generate surface electron waves [5, 27, 45, 79, 80]. When a biomolecule binds to the functionalized metal surface, it can be detected by observing the change in the propagation of the surface electron waves. Moreover, combining the SPR with fluorescence spectroscopy, or designing other SPR configurations, led to remarkable enhancement in the sensitivity [14, 37, 46, 54, 55, 68, 98].

Finding alternative approach to detect biological events is necessary as part of the process to find cheaper methods that can achieve the same or even higher sensitivities. One of these alternatives is field effect transistors (FETs) [50]. FETs are based on using traditional semiconductors, e.g., materials like metal oxides [37], organic semiconductors [30], silicon nanowires [12, 15], carbon nanotubes [2, 34, 65], or graphene [32, 71, 72, 85]. These systems are based on the change that takes place in the electronic structure due to the adsorption or binding of the biological molecules on the surface of their materials. While much progress has been made in improving the sensitivity of FET biosensors (indeed, sensitivity down to the nanomolar range has been demonstrated [65]), their limited sensitivity hampers the widespread application of these relatively simple and cheap FET-based sensors. The advances in graphene fabrication technologies and the possibility to grow large-area graphene sheets allowed it to be a promising candidate for several potential applications, in which one of them is biosensing.

1.2.1 Graphene Surface Modification

The first important step in developing a biosensor is adding functional groups in a controlled manner. Several reports are available on using graphene and graphene oxide as a biosensing platform [7, 16, 36, 38, 42, 51, 56, 70, 82, 86].

Surface biosensors are the class of biosensors that allows for detecting the target of interest on the surface without any contribution from a bulk signal, which helps in enhancing the sensitivity and the ability to perform the sensing in real time. Thus, adding functional groups in a controlled manner is a critical and essential step in developing biosensors. There have been several reports on graphene functionalization [6, 21, 22, 78]. Graphene oxide and exfoliated graphene are also widely used as a platform for sensing. Chemical vapor deposition (CVD) graphene is a more desirable platform because of the ease of fabrication, scalability, and its inherent low cost. In addition, CVD allows for transfer of graphene to any desired substrate.

In this work [39], we introduced the modification of single-layer graphene (grown via CVD) by assembly of a biotin moiety, which is functionalized with a diazonium salt (compound V, Scheme 1.1a). In order to prove that this layer can be used as a stable modification for biosensing, Streptavidin was bound to the biotin group, and the binding could be tested and monitored using a variety of optical techniques. Diazonium salts react with graphene via charge transfer complexation, with very little covalent functionalization [21].

The compound merely shifts the Dirac point, decreasing the sheet resistance because of charge transfer doping. The sheet resistance decreases by a factor of 2 (from 1200 Ω/sq to $\sim$600 Ω/sq) upon exposure of graphene to compound V. This mechanism is a critical advantage of this chemistry, as the optical and electronic properties are retained after modification. Compound IV was designed and synthesized (Scheme 1.1) where the amine end group can be diazotized and assembled on the graphene surface. Condensation of commercially available N-Boc-4-aminobenzyl alcohol (II) with biotin(I) in the presence of dicyclohexylcarbodiimide gave N-protected amino- benzyl ester of biotin(III) in 60% yield after purification as white needles. Compound III was deprotected

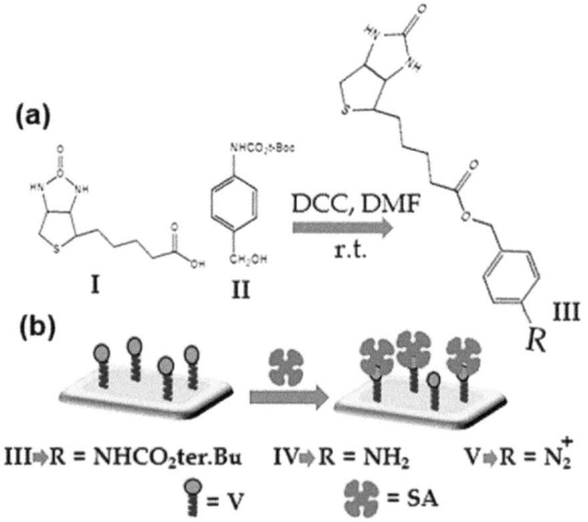

Scheme 1.1 (a) Synthetic scheme for compound V; (b) illustration of biotinylated graphene and the selective adsorption of Streptavidin used in this study. Reprinted with permission from Ref. [39], Copyright 2011, American Chemical Society.

by stirring a solution of III in dichloromethane containing five equivalents of trifluoroacetic acid. Evaporation of the solvent followed by treatment with sodium bicarbonate afforded arylamine (IV) as a light-yellow powder, which was used in the next steps without further purification. The synthetic route shown here is simple and generalizable to any desired sensing molecules, where the biotin can be replaced with any other desired molecule. The compound is thoroughly characterized and validated [39].

Graphene was prepared by CVD [53] using a copper foil annealed at 975°C in 6 sccm forming gas (5% H_2 in Ar) at a pressure of 500 mTorr for 10 min, followed by flowing ethylene for 10 min at the same temperature. The transfer method is described elsewhere [38]. The graphene was transferred to the substrate and annealed at 600°C in vacuum for 10 min in order to make sure that there is no remaining PMMA, which is used during the transfer, embedded in the graphene layer. Surface modification is achieved by the reaction of V with the graphene surface. For the surface modification, 10 mg of compound IV was suspended in 10 mL of acetonitrile mixed with

10 mg of nitrozonium tetrafluoroborate. The graphene was then functionalized with compound V by incubating the graphene in a solution of V, followed by washing with ethanol and water.

Compound V reacts with graphene via charge transfer complex formation where an electron is transferred from graphene to compound V. The biotinylated graphene is then used to detect SA in solution. SA was bound to the surface by incubating the modified graphene for 15 min, at room temperature, in a solution of 1 μM SA in PBS.

Several techniques were used to characterize the system described here. X-ray photoelectron spectroscopy (XPS) was performed using a monochromatic source, Al Kα ($h\upsilon$ = 1486.6 eV), and a photoelectron takeoff angle of 40°. The S2p photoelectron spectrum, consisting of a doublet peak at 164.3 and 165.5 eV, is shown upon functionalization of graphene with compound V. This is consistent with the presence of sulfur atom in the biotin moiety [29]. A low-intensity peak in the N1s photoelectron spectrum, which is consistent with previous reports on the reaction of diazonium salts with graphene, is also observed [83].

Raman spectroscopy measurements were performed using a Horiba ARAMIS Raman System with a HeNe laser source. The Raman spectra of single-layer graphene have three key characteristic peaks: the G peak, which is due to the vibrational mode of sp^2-bonded carbon; a 2D peak; and the D band, which is induced by "defects" in the structure. The difference between the spectra before and after modification is the dramatic increase in the D/G ratio after modification with V. The D/G ratio increases dramatically after modification with compound V [39]. This is in agreement with studies where donor and acceptor molecules are intercalated in few-layer graphene sheets [92]. The charge transfer complex and the resultant doping lead to an increase in the D/G ratio. Although no deterioration of the electrical properties is observed upon reaction with compound V, there may be some conversion of sp^2 to sp^3 carbon, which contributes to the D band, but it is not substantial enough to affect transport in these large-area films as evidenced by the decrease in sheet resistance following modification [21].

The surface modification was verified via real-time measurements of compound V binding to graphene using SPR spectroscopy.

SPR is one of the most powerful techniques used for the detection of biomolecules on surfaces [28, 102]. Surface plasmons are collective oscillations of the quasi-free electron gas at the interface between a metal and a dielectric. Absorption of molecules on the metal surface leads to a change in the refractive index and, consequently, a shift in the dispersion of the plasmon wavevector. This shift is used to determine information about the thickness and/or the surface density of the absorbed monolayer. In this work, a single layer of graphene, of the size 1 cm × 1 cm, was transferred to a 2 nm/50 nm chromium/gold film deposited on a glass coverslip. The sample was mounted on the SPR setup in a Kretschman configuration as described previously [45, 73, 79]. Compound V was reacted with the surface, in situ, using a 300 μL volume PDMS flow cell attached to the substrate. The binding was monitored in real time by measuring the change in the reflectivity of the gold substrate as a function of time. The difference between the reflectivity signal before binding and after washing with ethanol and water indicates the degree of modification (Fig. 1.1a). The binding of SA to the modified surface was also monitored by SPR. A 1 μM SA solution was bound to the surface; the real-time measurements of this binding are shown

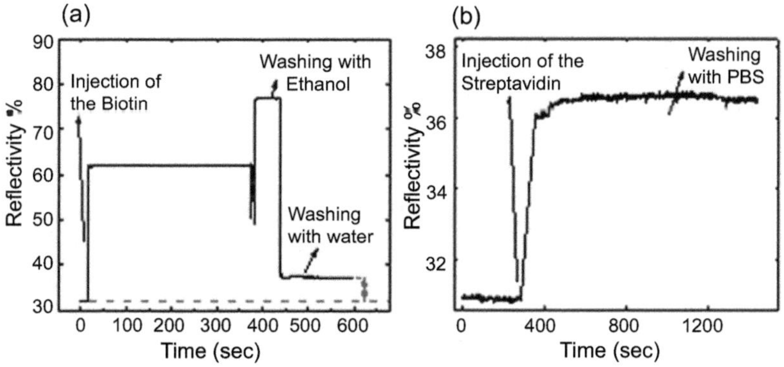

Figure 1.1 (a) Surface plasmon resonance (SPR) spectra showing real-time binding of V with graphene. The difference between the baseline and the signal after washing indicates the binding of compound V. (b) SPR showing real-time binding of Streptavidin to the biotinylated graphene. Streptavidin remains on the surface even after washing with PBS buffer. Reprinted with permission from Ref. [39], Copyright 2011, American Chemical Society.

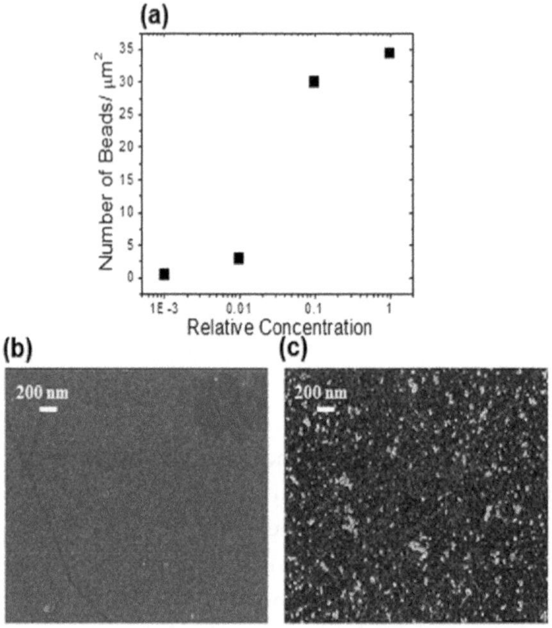

Figure 1.2 (a) Plot of the concentration of SA beads in solution versus the number of beads adsorbed to the surface of the biotinylated graphene as determined by SEM imaging. The plot shows a clear concentration dependence. (b) SEM image of bare graphene after incubation with the SA beads, followed by washing with water. This illustrates a lack of nonspecific binding of SA. (c) SEM image of the biotinylated graphene after incubation with the SA beads followed by washing with water. Reprinted with permission from Ref. [39], Copyright 2011, American Chemical Society.

in Fig. 1.1b. Upon washing with PBS, there is no change in the reflectivity signal due to the high affinity between biotin and SA. Reflectivity scans were performed after each binding step. Fitting of the reflectivity scans gave an SA surface density of 1.184 ng/mm^2. Scanning electron microscopy (SEM) imaging was used to determine the concentration dependence on SA binding. This was done by modifying the graphene substrate with the same concentration of compound V (40 mg/10 mL). The biotinylated graphene was then exposed to differing concentrations of SA-coated magnetic beads (Miltenyi Biotec estimated concentration of 1012 beads/mL) for 15 min. Figure 1.2a is a plot of the number of beads as a function of their concentration in solution, clearly indicating a concentration dependence. The reaction begins to saturate at a concentration above 10%.

One of the most critical issues in developing viable biosensors is preventing nonspecific binding of the biomolecules. Any nonspecific binding can contribute to the detected signal and give false positives. In our system, the graphene surface exhibited no detectable nonspecific binding when the unmodified graphene is exposed to SA. This was shown by incubating unmodified and biotinylated graphene layers in a solution of the SA-modified magnetic beads. The comparison is shown in Fig. 1.2b where the unmodified graphene shows no adsorption, while the biotinylated graphene is clearly modified with the beads (Fig. 1.2c). This indicates that graphene may not need to be passivated when used as a biosensor. Although one should not generalize to all biomolecules, this could provide a distinct advantage over traditional platforms such as gold films where nonspecific adsorption is difficult to prevent. In summary, a compound was designed and synthesized to biotinylate graphene using a diazonium salt functionality. The surface modification was verified using XPS, Raman spectroscopy, SPR spectroscopy, and SEM imaging. Importantly, the chemistry is benign and does not destroy graphene's unique electrical and optical properties. The biotinylated graphene was shown to effectively sense SA in solution, and the adsorption was measured in real time using SPR spectroscopy. One important observation is that the unmodified graphene surface does not allow nonspecific binding as demonstrated in our system. This provides a distinct advantage of using graphene as a sensing platform, as compared to traditional substrates.

1.2.2 Graphene as a Strong Fluorescence Quencher

Fluorescence quenching is a powerful technique used to obtain information about the dynamic changes in proteins in complex macromolecular systems. In this work, graphene is shown to be a very efficient quencher of fluorescence molecules where the quenching effect was one order of magnitude higher than that of gold. The fluorescence intensity was distance dependent where increasing the distance between the fluorescence molecule and the graphene surface from 4 to 7 nm increased the fluorescence intensity by a factor of 7.5. This type of distance dependence suggests a nonradiative nature in the energy transfer between the graphene and the fluorophore due to the excitation of an exciton.

Fluorescence quenching is one of the most powerful techniques for obtaining information about the conformational or dynamic changes in proteins in complex macromolecular systems [59, 103]. The emission of a fluorophore is perturbed by a quencher, which can be a substrate or another molecule, under certain conditions that allow for energy transfer between the fluorophore and the quencher. Near a metal surface, the emission of the excited fluorophore is altered by interaction with the electric field at the metal surface. The interactions between metal surfaces and fluorophores through radiative or nonradiative decay are well-understood [9, 77]. Graphene is a unique interface for fluorescence energy transfer because it is a nearly transparent semimetal. Graphene has a linear band dispersion around the corners of its Brillouin zone [67] and a nearly constant optical absorption, and thus optically excited species can be quenched by resonant energy transfer via excitation of electron–hole pairs in the graphene. This behavior was observed between semiconducting nanocrystals and graphene [8]. The rate of resonance energy transfer from the excited state of a dye to the π system of graphene was also theoretically studied [87, 88].

In this work [41], graphene is shown to be a very efficient quencher for fluorescent molecules, as compared with metals. The fluorescence quenching is distance dependent, which supports the nonradiative nature of the energy transfer between the graphene and the fluorophores. This effect can, potentially, be used to study protein–protein interactions and dynamic changes in biomolecules.

Growth and transfer were performed as previously explained. Graphene was transferred to several substrates, including gold, quartz, silicon, and 300 nm SiO_2 on Si depending on the type of measurement that was performed. Fluorescence-labeled molecules were attached to graphene through biotinylated compounds of varying lengths (Scheme 1.2). The synthesis of compound I and its binding to graphene is explained elsewhere [54, 90]. SA labeled with Cy5 dye (Invitrogen) was bound to the modified graphene by incubating the substrate for 15 min at room temperature in a

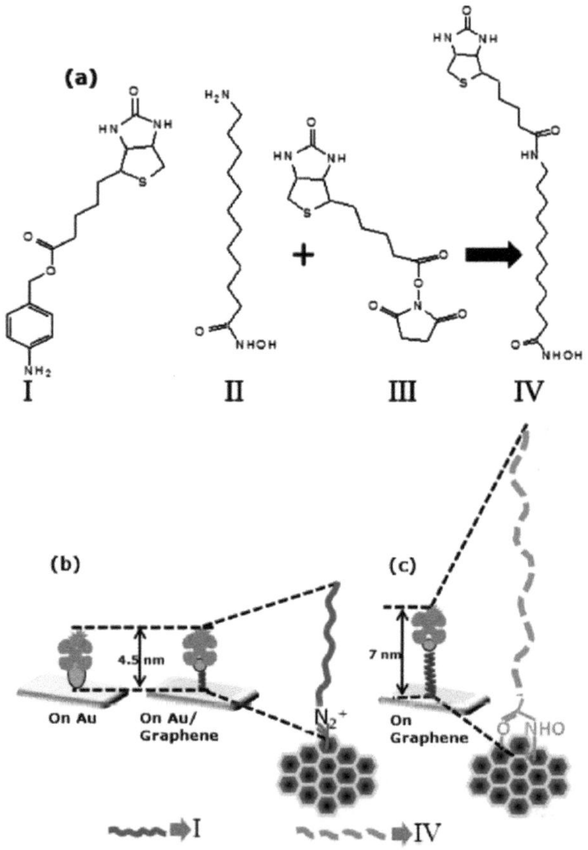

Scheme 1.2 (a) Compounds used for self-assembly on graphene. Compound I was converted to a diazonium salt prior to reacting with graphene. The reaction of compounds II and III yields hydroxamic acid IV. (b, c) Diagrams showing the self-assembly of different-chain-length compounds on graphene and gold. Reprinted with permission from Ref. [39], Copyright 2011, American Chemical Society.

solution of the modified Cy5 dye. This modification yields a modified graphene substrate where the labeled SA is 4 nm above the substrate as measured by SPR. This distance was increased by modifying graphene with compound IV. Compound IV was assembled on graphene by incubating overnight and then washing with ethanol to remove the unbound compound. NHS-Biotin(III) was applied to the surface overnight, followed by washing with dimethylformamide (DMF) and ethanol, and SA labeled with Alexa Fluor dye was applied for 15 min; the longer spacer achieves a distance of ~7 nm after the SA binding.

For ellipsometry measurements, a J. A. Woollam VUV-VASE ellipsometer VU-302 was used to determine the optical constants (refractive index *n* and extinction coefficient *k*) of a single-layer graphene transferred to a quartz substrate. Graphene samples were measured in the wavelength range of 150–800 nm, and the data (with backside correction applied for quartz substrates) were fitted to obtain the *n* and *k* values by building a general oscillator model. The thickness of the graphene layer was determined by medium energy ion scattering (MEIS) measurements [11]. The graphene layer was transferred to an Si substrate, and the MEIS measurements were performed using an HVEE system (100 KeV protons).

Backscattered protons were analyzed with a toroidal electrostatic energy analyzer [10, 47]. SPR and surface plasmon fluorescence spectroscopy (SPFS) measurements were done using a home-built SPR setup using a 633 HeNe laser. The sample was mounted on the SPR setup in a Kretschman configuration [13, 45, 47, 73, 79, 84, 90]. The biotin-based monolayer was assembled on the surface in situ using a 300 μL volume PDMS flow cell attached to the substrate. Binding of the SA to the surface was also monitored by SPR, where 1 μM SA was applied in situ to the surface. A reflectivity and fluorescence scan was performed after each layer was deposited. Fitting of the SPR data to determine the distance above the surface was done using Winspall software using a model based on the Fresnel and Maxwell equations. Fluorescence emission was also measured by direct excitation of the fluorophores using an ARAMIS Raman System from Horiba at 532 nm excitation wavelength.

To study the effect of graphene on fluorescence intensity, we excited fluorophores in the total internal reflection mode using SPFS and by direct excitation using a spectrometer equipped with a 532 nm laser source and a cooled charge coupled device detector. SPFS utilizes the evanescent electromagnetic field of a surface plasmon to excite fluorophores near the surface [37, 47, 54, 84, 90]. Graphene was transferred to a glass substrate coated with 2 nm Cr/50 nm Au to use the plasmon mode in the gold layer. For comparison, similar measurements were performed on bare gold. Nonspecific binding to a bare graphene surface does not occur as previously demonstrated [39], so the collected fluorescence signal was only from the bound species at the defined distance and not due to any false signal from nonspecific binding.

There are two key parameters in this experiment: the distance between the metal and the fluorescence molecule and the surface density of the fluorescence molecules. Both should be accurately calculated to eliminate the quenching effect of the metal layer and measure only the quenching effect of graphene.

To calculate the distance and the surface density, SPR reflectivity scans were performed and fitted by a model based on the Fresnel equations. To determine the fitting parameters, the thickness and the optical constants (n and k) of the graphene layer were obtained. The graphene thickness was determined from the MEIS measurements [11]. In this system, protons with energy of 100 keV are incident along a major crystallographic direction in the solid. Energy- and angle-resolved detection of backscattered ions provides surface structural and compositional information. MEIS measurements indicated a graphene thickness of 0.4 nm. Refractive index and extinction coefficient (n and k) were measured by ellipsometry. Fitting the ellipsometer results gave the n and k values for the graphene layer of 2.7 and 2, respectively, at 630 nm. The diazotized compound I (Scheme 1.2a) was first bound to the graphene surface, and the binding was monitored in real time, followed by binding of the SA labeled with Cy5 dye. The binding of the SA was also monitored in real time by measuring the fluorescence signal. Reflectivity and fluorescence scans were performed after each binding step. The results obtained from MEIS and ellipsometry were used to fit the SPR reflectivity scan data to determine the thickness of each layer. After the biotin compound and labeled SA were applied to the graphene surface, the distance above the metal was found to be 4.5 nm, which includes the graphene layer. To apply a similar system above the gold layer, we used Anti-SA to be directly physisorbed to the gold surface, and then Cy5-labeled SA was bound to it; this system gave a similar distance of 4.5 nm. The surface density was determined to be 1.184 and 2.122 ng/mm^2

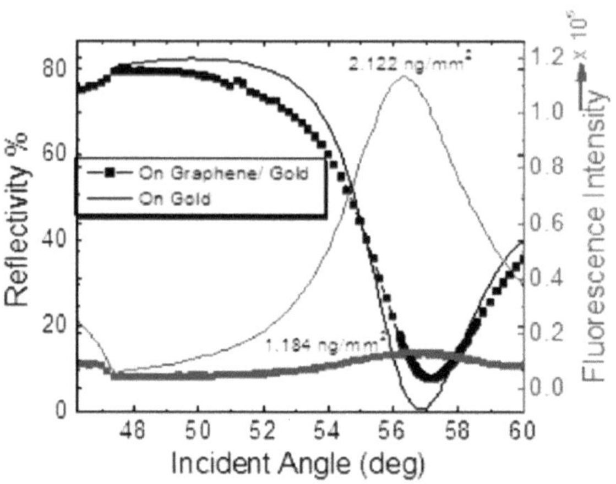

Figure 1.3 Reflectivity and fluorescence scans as a function of the incident angle of both gold and graphene on gold after binding of the Cy5-labeled Streptavidin to the linker attached to both surfaces. The graphs show a reduction in the fluorescence intensity on the graphene surface due to its quenching effect. The surface densities in both cases calculated from the angle shift were found to be 1.184 ng/mm^2 on graphene and 2.122 ng/mm^2 on gold. Reprinted with permission from Ref. [41], Copyright 2012, American Chemical Society.

on both graphene and gold, respectively. The surface densities were calculated from the SPR shift [98].

Fluorescence scans as a function of the incident angle were performed after applying the Cy5-SA layer in both cases. The results are shown in Fig. 1.3. The fluorescence intensity on the graphene/gold system is almost one order of magnitude less than that on gold despite having the fluorescence molecule at the same distance above both substrates. Additionally, the surface density is 1.8 times less than that on gold. Both measurements were performed under the same conditions of laser intensity and binding time. This large reduction in the fluorescence intensity clearly indicates the highly efficient fluorescence quenching of graphene. Anomalously high efficiency quenching was also observed with semiconducting nanocrystals on graphene [67]. At a distance of 4 nm above graphene, one cannot assume the quenching is due to charge transfer where the charge transfer takes place at very small distances [69]. Another possible mechanism is the nonradiative decay of the fluorescence molecule via creation of an electron–hole pair [77, 95]. When fluorophores are in close proximity to metals, the metal layer alters the way an excited fluorophore loses its energy [4, 54, 90]. Additional decay channels contribute to the decrease in the radiative quantum yield of the fluorophores. They take place at different characteristic fluorophore–metal distances.

At a distance less than 10 nm of the metal surface, the nonradiative decay of fluorescence is the dominating process. The excitation is assumed to be due to dipole–dipole interactions arising from excited electron–hole pairs (excitons) in the metal. The energy-transfer model (Förster model) [48] gives an R^{-6} dependence of the transfer rate to the separation distance due to the co-effect from both near fields of the donor molecule and the acceptor molecule. However, the distance dependence of energy transfer can be greatly compromised to R^{-3}–R^{-4} due to the integration over the enlarged number of effective acceptor sites. At an intermediate distance regime (a few nanometers up to ~20 nm), a significant fraction of the excited fluorescence couples back to surface plasmon polaritons of the metal by fulfilling momentum matching conditions, and it represents a significant loss of fluorescence yield. At sufficient separation distances (>20 nm), free emission of the fluorophore dominates. These kinds of interactions between a fluorophore and a metal are theoretically calculated and quantitatively measured [91]. To compare these effects of metals on fluorescence emission to that of graphene, we measured the fluorophore at different distances above a single graphene layer transferred to 300 nm SiO_2 on Si substrates and measured the fluorescence intensity by direct excitation. In this case, SA labeled with Alexa Fluor 532 was used. To increase the distance, we applied an 11-aminoundecylhydroxamic acid molecule overnight to the surface; NHS-biotin was bound by incubating overnight, followed by washing with DMF and ethanol; then, labeled SA was applied. Both systems should achieve an estimated total distance of 4 and 7 nm, respectively. Applying the hydroxamic acid to metal oxides forms a chain tilt of 20° [23] where the distance between the fluorophore and the graphene is ~6.8 nm.

The results of fluorescence intensity above the graphene at different distances are shown in Fig. 1.4. These results show an order

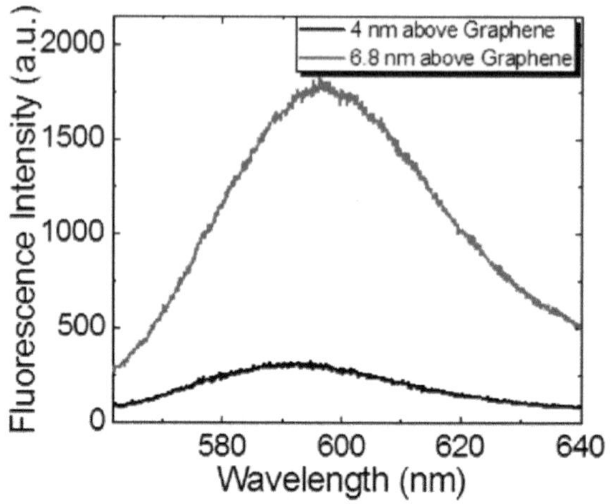

Figure 1.4 Fluorescence intensity measured by direct excitation showing the increase by a factor of 7.5 with increasing the distance above the graphene from 4 to 6.8 nm. Reprinted with permission from Ref. [41], Copyright 2012, American Chemical Society.

of magnitude decrease due to doping [87, 88], which translates to a decrease in the fluorescence intensity. This point still needs further investigation.

1.3 Graphene as a Transparent Electrode

Transparent conducting electrodes (TCEs) constitute a crucial component of photovoltaic and display technologies. Candidate materials have to be (1) highly transparent and (2) of low sheet resistance. The widely used indium tin oxide (ITO) suffers from its high cost and brittleness. Carbon-based materials, such as carbon nanotubes and graphene, have been recently considered TCE materials [61, 66, 74].

A single-layer graphene sheet has a transparency of about 98% [33]. The high transparency of graphene (98%) nominates it as a TCE material, but its relatively high sheet resistance necessitates its doping. Graphene doping is, in general, unstable, primarily because of the relative chemical inertness of graphene arising from its

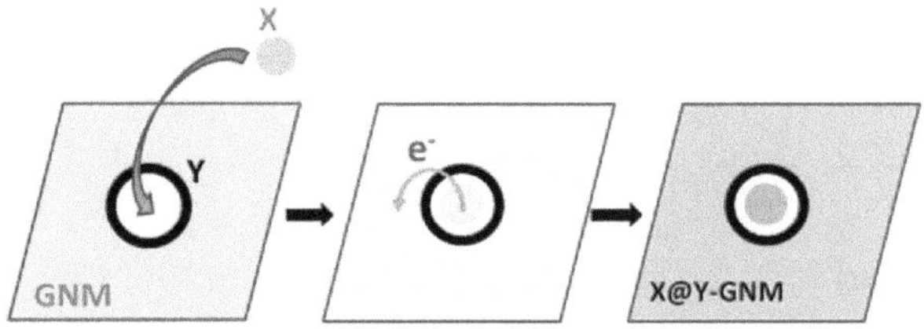

Figure 1.5 A scheme showing the possibility of the charge transfer by introducing a neutral dopant to the graphene nanopore. Reprinted with permission from Ref. [60], Copyright 2013, American Chemical Society.

sp^2 bonding. Dopants physisorb on its surface, which renders the lowering of the resistance only temporary. Without a stable doping method for graphene, its use as a TCE will be greatly limited.

1.3.1 Graphene Nanomeshes as TCE Materials

GNMs come to the rescue by offering controlled chemistry through their pores. Passivation of the pore edges by a species different from carbon results in the creation of edge dipoles. This allows for the controlled, stable, chemical doping of GNMs [75]. A neutral moiety may react with the GNM, causing a charge transfer reaction that leads to the ionization of this moiety. The local dipole moments at the pore edge electrostatically trap the resulting ion. Doping the GNM moves the Fermi level into the valence or the conduction bands (Fig. 1.5).

The first-principle calculations show that the band structure of the GNM is not altered by the doping process. Furthermore, the dopant state involved in the charge transfer is located away from the Fermi level and hence is not expected to cause electronic scattering in a GNM-based device. This doping approach, termed chelation doping, offers a controlled, stable, and rigid-band doping of GNMs [60].

Pore edge dipoles are formed by passivating the GNM pore with chemical compounds different from carbon. These dipoles provide the electrostatic trap for the dopants. Figure 1.6a shows a hydrogen- and an oxygen-passivated GNM (H-GNM and O-GNM, respectively), where all chemical bonds are satisfied, which forms a chemically

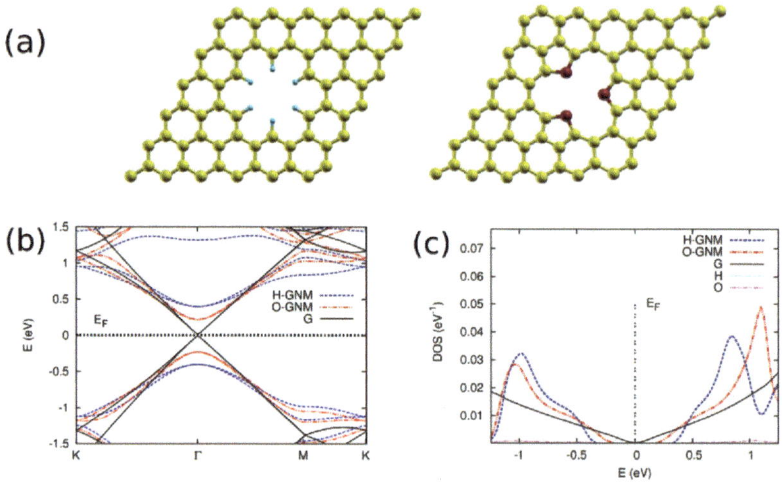

Figure 1.6 (a) A hydrogen- and an oxygen-passivated GNM (H-GNM and O-GNM). (b) The band structures of the two passivated systems and pristine graphene. (c) The density of states (DOS) of H-GNM, O-GNM, and graphene. Reprinted with permission from Ref. [60], Copyright 2013, American Chemical Society.

stable GNM. The band structures of the two passivated systems and pristine graphene are shown in Fig. 1.6b. The first-principle calculations predict a gap of 0.7 eV for the H-GNM and 0.4 eV for the O-GNM. Away from the gap region, the bands are linear, with a group velocity that is half of graphene's. The density of states (DOS) of H-GNM, O-GNM, and graphene are shown in Fig. 1.6c. The projected DOS (PDOS) shows that the passivating species do not have any significant contribution in the linear region or at the band edges and are, therefore, not expected to cause any electronic scattering [94].

For n-doping, passivating the pore edge with oxygen leads to the creation of pore edge dipoles due to the electronegativity mismatch between carbon and oxygen. This allows for hosting positive ions in the pore region. Electron-donating elements (e.g., lithium, sodium, and potassium) can be used as dopants. A potassium atom brought to the vicinity of the pore donates its 4s electron to the GNM carbon atoms and thus dopes the GNM (Fig. 1.7a). The doping is ultra-stable, with an LDA-predicted binding energy of 2.8 eV. A comparison of the

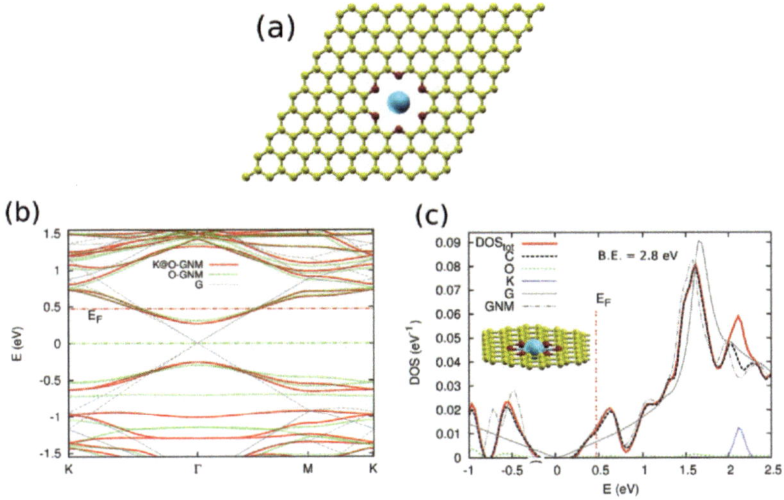

Figure 1.7 (a) Doping the O-GNM with a potassium atom by donating its 4s electron to the O-GNM skeleton. (b) The band structures of the doped and undoped systems indicating that the potassium chelation preserves the O-GNM band structure. (c) The density of states of the potassium-doped O-GNM showing contributions of different species. The undoped O-GNM and pristine graphene are included for comparison. Reprinted with permission from Ref. [60], Copyright 2013, American Chemical Society.

band structures of the doped and undoped systems indicates that the potassium chelation does not alter the GNM band structure.

As we see in Fig. 1.7b, the effect of the potassium atom is to shift the Fermi level into the conduction band. The DOS and PDOS of the potassium-doped GNM confirm this doping picture (Fig. 1.7c). The location of the potassium empty 4*s* state is far from the Fermi energy and hence is not expected to affect the transport properties of the doped GNM.

A p-doped GNM system is formed by chelating an electron-*accepting* species in a hydrogen-passivated pore. A fluorine atom brought close to the pore of the GNM leads to the GNM transferring one of its electrons to the fluorine atom, thus doping the GNM and ionizing the fluorine (Fig. 1.8a). The fluorine ion docks in the pore, by electrostatically binding to the pore edge dipoles. The GNM is now p-doped. The band structures of the doped and undoped

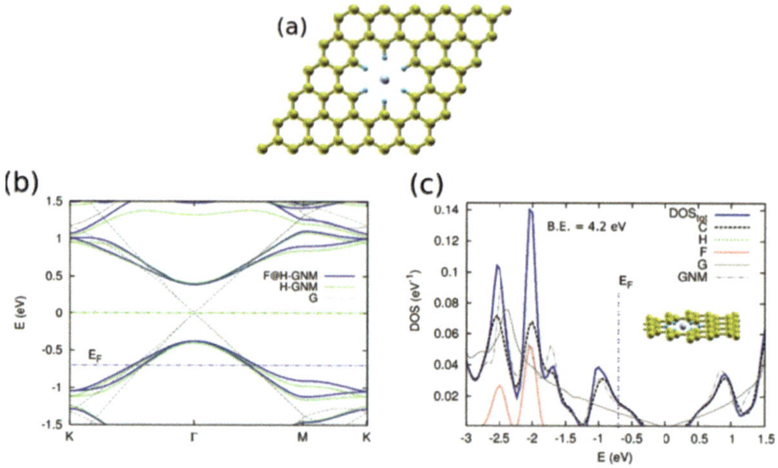

Figure 1.8 (a) H-GNM doped with fluorine atom. (b) The band structures of the fluorine-doped and undoped systems. (c) The total density of the fluorine-doped H-GNM showing the contributions of different species. The undoped H-GNM and pristine graphene are included for comparison. Reprinted with permission from Ref. [60], Copyright 2013, American Chemical Society.

systems are plotted. Figure 1.8b shows that the doping causes no significant change in the band curvatures, and the Fermi level of the doped system is in the linear region of the valence band, indicating that chelation p-doping of GNMs occurs in a rigid-band way.

This rigid-band picture is confirmed by inspecting the DOS of the doped H-GNM system. Figure 1.8c shows the total DOS, and the PDOS. The Fermi level of the fluorine-doped H-GNM system indicates that the GNM is p-doped. The doped and undoped systems have very similar DOS, except for the fluorine-occupied 2p states on the left side of the gap. These states are 1.3 eV away from the Fermi level and are, therefore, too far to obstruct low-energy electronic transport.

Therefore, ion chelation doping offers a novel way to stably dope graphene structures, making it possible to utilize the exceptional electronic and optical properties of graphene and its derivative structures in various technological applications.

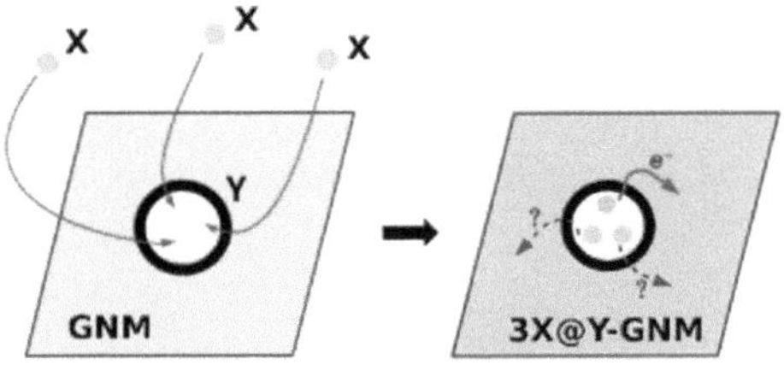

Figure 1.9 A conceptual sketch of multiple-ion chelation doping of a GNM. Chelant atoms are successively brought close to pore, where some or all of them become ionized, thus doping the GNM. Here, n-doping is shown, p-doping simply replaces the electron with a hole. In principle, partial charge transfer can occur, where the highest occupied molecular orbital (HOMO)/lowest unoccupied molecular orbital (LUMO) of the chelants can move relative to the GNM spectrum. Reprinted from Ref. [18], Copyright 2018, with permission from Elsevier.

Fabrication techniques are likely to yield GNMs with nanometer-sized pores, which can host more than one dopant atom. This opens the door for some important questions regarding the dopant stability in the pores. How would multiple dopants behave in a nanometer-sized pore? Would the dopants uniformly distribute around the pore edge, or would they cluster together, with obvious impact on their binding with the GNM? In addition, how would the distribution/clustering of the dopants in the pore affect the rigid-band doping picture?

In GNMs with nanometer-sized pores, electrostatic dopant binding is expected to decrease. This is due to the following reasons: (1) the increase in the charge transferred to the sheet and hence its energy cost; (2) the repulsion between ionized dopants; (3) the weaker interaction between a dopant and the pore dipoles; (4) dopants may cluster in large pores. This implies that chemical binding, well known in metals, may occur between the dopant atoms, which may alter the doping physics occurring with single dopants [18] (Fig. 1.9).

We begin by presenting the electronic properties of our undoped large-pore GNM systems. Figures 1.10a,b show an oxygen- and a hydrogen-passivated GNM (insets), with supercell size of 9×9 graphene unit cells (O-GNM9 and H-GNM9), as well as their DOS. Both GNMs have a pore size of about 0.8 nm and are chemically

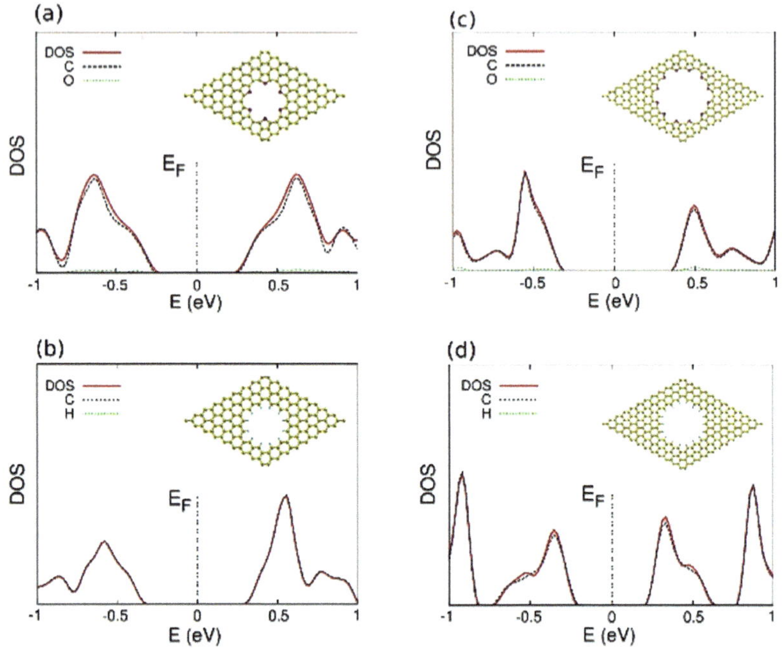

Figure 1.10 DOS and PDOS of the pristine (a) O-GNM9, (b) H-GNM9, (c) O-GNM12, and (d) H-GNM12 systems. The insets show the unit cells of the relaxed structures. Reprinted from Ref. [18], Copyright 2018, with permission from Elsevier.

stable as the passivation saturates all bonds. The two GNMs are semiconducting, with a gap of about 0.5 eV for O-GNM9, and 0.7 eV for H-GNM9. The PDOS shows that the states of the passivating species are located far from the top/bottom of the valence/conduction bands and are, therefore, not expected to induce resonant scattering. Two GNMs with larger pores are shown in Figs. 1.10c,d. The oxygen-passivated GNM, O-GNM12, has a pore diameter of 1.6 nm, while the hydrogen-passivated GNM, H-GNM12, has a pore diameter of 1.3 nm. Both systems are intrinsically semiconducting, with gap sizes of 0.8 eV and 0.4 eV for the oxygen-passivated and hydrogen-passivated systems, respectively.

We first discuss the n-doping cases. A lithium atom brought close to the edge of the pore of O-GNM9 (inset of Fig. 1.11a) loses its electron to the graphene lattice, thereby getting ionized and

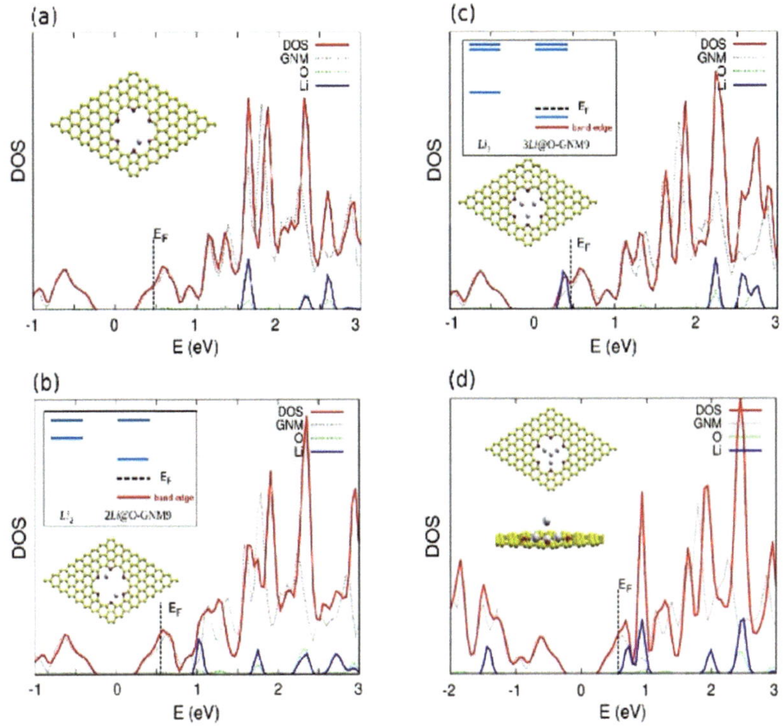

Figure 1.11 DOS of the (a) Li@O-GNM9, (b) 2Li@O-GNM9, (c) 3Li@O-GNM9, and (d) 4Li@O-GNM9 systems. The bottom insets refer to the relaxed structures. In (b) and (c), the top insets show the Li states within the corresponding O-GNM9 system, as compared to those of the isolated Li clusters. The DOS of the pristine O-GNM9 (GNM) is shown for comparison. Reprinted from Ref. [18], Copyright 2018, with permission from Elsevier.

becoming electrostatically trapped in the field of the pore dipoles, forming an Li@O-GNM9 structure. The average Li–O distance is 2.19 Å, and the binding energy is 1.29 eV. This is significantly higher than the adsorption energy of Li atop pristine graphene, 0.4 eV, as shown by recent studies [58]. This is due to the electrostatic interaction between the Li and the dipoles formed by the C–O bonds of the pore edge.

The DOS and PDOS of the Li@O-GNM9 system are shown in Fig. 1.11a. The Fermi level location indicates that the O-GNM9 is n-doped. The O-GNM9 perturbs the Li state such that it raises its

2s state by about 1.1 eV above the Fermi level. The Li 2s state is, therefore, too far up in the spectrum to obstruct low-energy electronic transport. By integrating the DOS from the conduction band edge to the Fermi level, we find that one electron has been transferred to the O-GNM9.

With two Li atoms, the second Li settles at the pore edge opposite to the first one (inset of Fig. 1.11b), with an Li–Li separation of 5.28 Å, and in the same plane of the GNM. The binding energy of the two lithium atoms in the pore is found to be 0.79 eV. This in-pore configuration binding energy is about 0.50 eV higher than that of the system with the second lithium atom atop a graphene skeleton, whether far from the pore, or close to it. The DOS and PDOS of the 2Li@O-GNM9 system are shown in Fig. 1.11b. The shift in the Fermi energy indicates the increase in the doping level of the 2Li@O-GNM9 over the Li@O-GNM9 system. Calculations confirm a charge transfer of 2e between the two lithium atoms and the graphene lattice. Thus, the two Li atoms double the doping level and maintain the rigid-band doping picture, but with a smaller binding energy.

Adding a third lithium atom to the pore causes a qualitative change. While the states of the single and double Li-doping are far above the conduction band edge of O-GNM9, one of the states of the 3Li cluster falls in the conduction band of O-GNM9 (Fig. 1.11c). The O-GNM9 perturbs the Li and 2Li spectra in such a way to raise their atomic/molecular HOMO states, leading to the ionization of lithium atoms in both cases. In the 3Li case, the perturbation is such that the lowest lithium 2s-like state lies in the conduction band of the system below the Fermi level, and therefore two electrons occupy that state, while the third electron is donated to O-GNM9, thus doping it to a level similar to that of the Li@O-GNM9 system. The total binding energy in this case is 1.91 eV.

The fourth Li atom of the 4Li@O-GNM9 system sits about 2.79 Å above the plane of the GNM (Fig. 1.11d). The 4Li@O-GNM9 system has the lowest Li_4 state *below* the valence band edge, with *two* electrons transferred to O-GNM. The binding energy of the 4Li@O-GNM9 is 1.59 eV. At higher doping (5Li@O-GNM9 and 6Li@O-GNM9), an Li state persists in the conduction band of the system.

The situation for Na-doping case is qualitatively similar to that of Li-doping. Figure 1.12 shows the structure and DOS of Na@O-

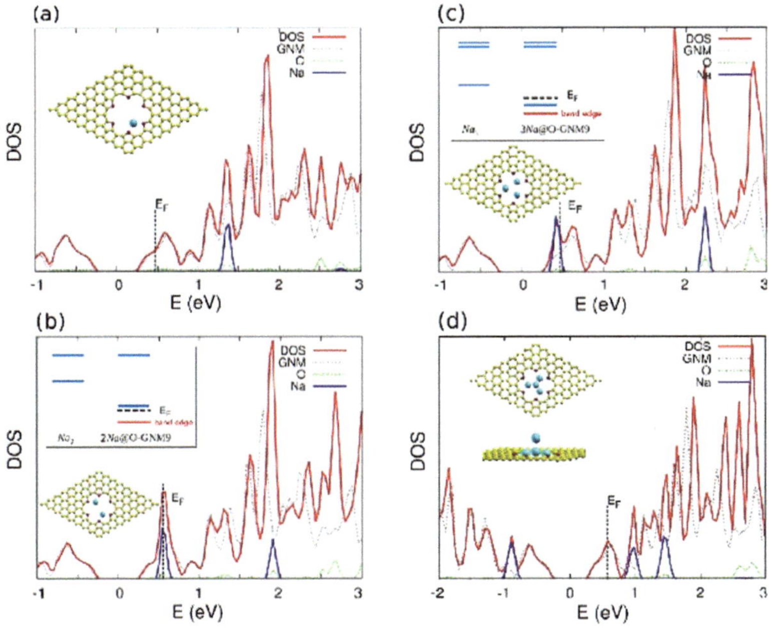

Figure 1.12 DOS of the (a) Na@O-GNM9, (b) 2Na@O-GNM9, (c) 3Na@O-GNM9, and (d) 4Na@O-GNM9 systems. The insets refer to the relaxed structures. In (b) and (c), the top insets show the Na states within the corresponding O-GNM9 system, as compared to those of the isolated Na clusters. The DOS of the pristine O-GNM9 (GNM) is shown for comparison. Reprinted from Ref. [18], Copyright 2018, with permission from Elsevier.

GNM9, 2Na@O-GNM9, 3Na@O-GNM9, and 4Na@O-GNM9 systems. In the case of the Na@O-GNM9 and 2Na@O-GNM9 systems, the 3s Na states are located above the Fermi level, and the systems are n-doped with a charge transfer of 1 and 2 electrons, respectively. Their binding energies are 1.62 eV (significantly higher than that of an Na atom atop pristine graphene [75]), and 1.57 eV, respectively. For the 3Na@O-GNM9 system, the binding energy is 2.79 eV, and the lowest Na3 state lies in the conduction band below the Fermi energy, hosting two of the three Na3 electrons, while the third electron is transferred to the GNM. The corresponding state of the 4Na@O-GNM9 system is pushed down further to the valence band.

We now turn to n-doping of the larger pore, where we consider six systems: Li@O-GNM12, 2Li@O-GNM12, 2Na@O-GNM12, 3Li@O-

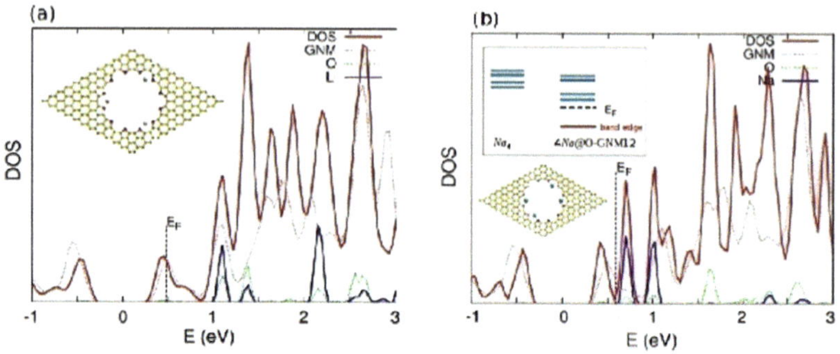

Figure 1.13 DOS of the (a) 3Li@O-GNM12 and (b) 4Na@O-GNM12 systems. The bottom insets refer to the relaxed structures. In (b), the top inset shows the Na states within the corresponding O-GNM12 system, as compared to those of the isolated Na4 cluster. The DOS of the pristine O-GNM12 (GNM) is shown for comparison. Reprinted from Ref. [18], Copyright 2018, with permission from Elsevier.

GNM12, 4Na@O-GNM12, and 6Na@O-GNM12. For 1 and 2 dopant atoms, the behavior is qualitatively similar to the small-pore case. For 3 and 4 dopants, the situation is different. Figure 1.13a shows the structure and DOS of the 3Li@O-GNM12 system, indicating that all three Li atoms are now ionized, with three electrons transferred to the GNM. The system has a binding energy of 1.39 eV, with its three Li-like states above the Fermi energy. The 4Na@O-GNM12 case is shown in Fig. 1.13b. All Na atoms are ionized, contrary to the small-pore 4Na@O-GNM9 system, with four electrons now doping the GNM. The O-GNM12 pulls down the lowest two states of the Na4 cluster, positioning them just above the Fermi level of the system, signaling that higher loads will result in pushing Na states down into the conduction band of the GNM. Indeed, with six Na atoms, the 6Na@O-GNM12 system has its lowest Na state below the Fermi energy, now with only four electrons doping the GNM.

We now examine the p-doping of large-pore GNM systems. We consider six doped H-GNM9 systems (pore size 0.8 nm), and 4-doped H-GNM12 systems (pore size 1.3 nm), using F and Cl. The reported binding energies are calculated from the energies of the most stable form of the dopants (F2 and Cl2 molecules).

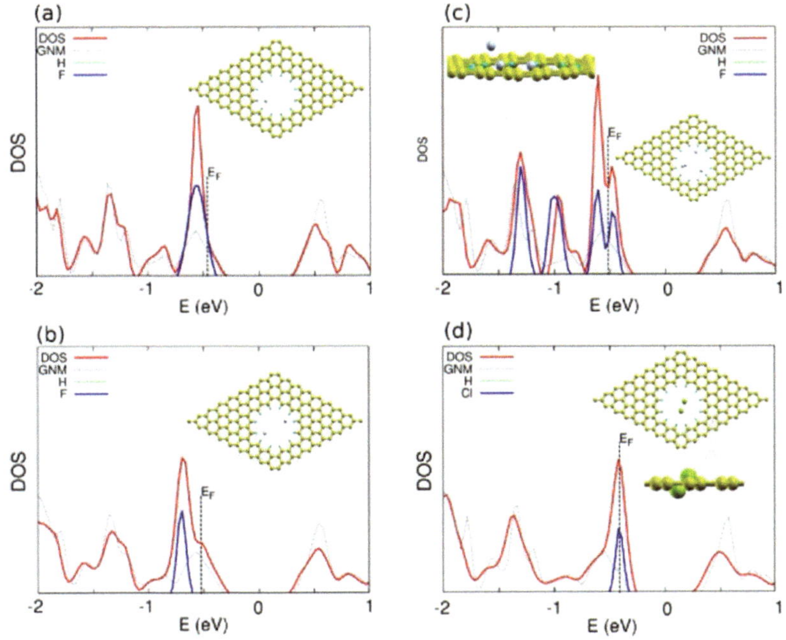

Figure 1.14 DOS of the (a) F@H-GNM9, (b) 2F@H-GNM9, (c) 4F@H-GNM9, and (d) 2Cl@H-GNM9 systems. The inset figures refer to the relaxed structures. In (c), the inset shows a side view of the 4F@H-GNM9 system, indicating the off-plane structure of the 4F dopants. The DOS of the pristine H-GNM9 (GNM) is shown for comparison. Reprinted from Ref. [18], Copyright 2018, with permission from Elsevier.

A fluorine atom brought close to the edge of the H-GNM9 pore will have its spectrum perturbed by the pore dipoles, resulting in the LUMO of the F falling below the valence band edge of the GNM. This will cause an electron to spill from the GNM skeleton to occupy the F LUMO, thus ionizing it and p-doping the GNM. The electrostatic attraction with the nearby hydrogen atoms creates an ultra-stable environment for the F anion. Figure 1.14a shows the structure and DOS of the F@H-GNM9. The average F–H distance with the nearest two hydrogen atoms is 1.70 Å. The binding energy of F@H-GNM9 is found to be 1.07 eV. The H-GNM causes the F $2p_z$ state to be pinned about 0.3 eV below the valence band edge of H-GNM, and 0.1 eV below the Fermi level. The structure and DOS of the 2F@H-GNM9 system is shown in Fig. 1.14b. The GNM is p-doped, with the Fermi energy located 0.3 eV below the valence band edge. The binding energy of the system is 1.72 eV. The two F atoms are now ionized, with their $2p_z$ states full and located 0.2 eV below the Fermi level. Doping with three fluorine atoms leads to a qualitatively similar scenario.

To reach the maximum pore capacity of H-GNM9, we add a fourth fluorine atom to form 4F@H-GNM9. The system reaches its minimum energy by having two of its fluorine atoms form an elongated fluorine molecule, F_2^*, which protrudes out of the pore plane (Fig. 1.14c), with a bond length of 1.68 Å (an F_2 molecule has a bond length of 1.43 Å). The F_2^* is 2.21 Å from the closest pore edge hydrogen. The DOS of the 4F@H-GNM9 system shows that there is one fluorine state just above the Fermi energy. Inspection of this state indicates that it is a superposition of the $2p_z$ states of the two fluorine atoms forming the elongated molecule. These two fluorine atoms are not ionized. This is confirmed by the Lowdin charge analysis, which also shows that the other two fluorine atoms are each singly ionized. The binding energy of the 4F@H-GNM9 system is 2.02 eV, which is only ~0.10 eV higher than that of the 3F@H-GNM9 system, suggesting that 3F@H-GNM9 is the maximally doped and most stable fluorine-doped H-GNM9 structure, and hence best suited for a p-doped H-GNM9-based device. An energy scan of the 3F@H-GNM9 system where the fluorine dopants are moved above the pore indicates that no other minima exist in the pore vicinity.

Next we describe our p-doping results for larger pores. We form a hydrogen-doped GNM with a pore size of about 1.3 nm, H-GNM12. Since the pore is large and three fluorine atoms can stably dope the smaller pore system, we begin with 3F@H-GNM12 (Fig. 1.15a). The average distance between the fluorine and the closest two hydrogen atoms in the structurally relaxed system is 1.67 Å.

The average FeF distance is 8.62 Å. The DOS of the system shows that the GNM12 is p-doped, with the Fermi energy placed 0.1 eV below the valence band edge. The fluorine $2p_z$ states are all full and are located at least 0.3 eV below the Fermi level. The integrated DOS shows that the GNM lost three electrons. The binding energy of the 3F@H-GNM12 is 2.72 eV.

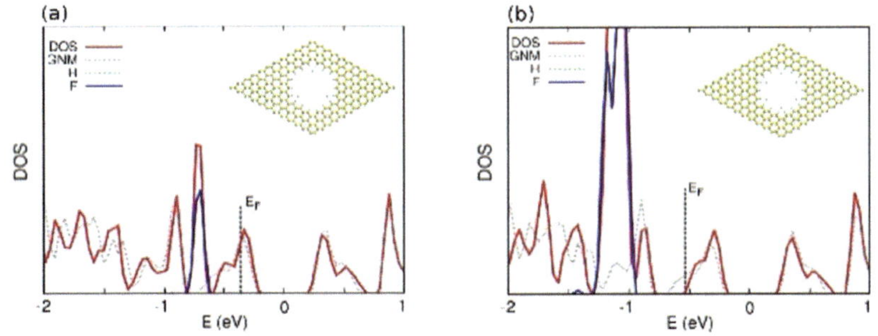

Figure 1.15 DOS of the (a) 3F@H-GNM12 and (b) 6F@H-GNM12 systems. The inset figures show the relaxed structures. The DOS of the pristine H-GNM12 (GNM) is shown for comparison. Reprinted from Ref. [18], Copyright 2018, with permission from Elsevier.

Adding more fluorine atoms leads to similar physics. Doping with four and six fluorine atoms removes four and six electrons from the GNM, respectively, confirming the full ionization of the fluorine atoms in both systems. We only show the 6F@H-GNM12 system here (Fig. 1.15b). The highest filled fluorine $2p_z$ state is 0.5 eV away from the Fermi level, which is pinned 0.3 eV below the valence band edge. The binding energy of the 6F@H-GNM12 system is 2.76 eV. Finally, Cl doping of the larger pores gives slightly more stable structures compared to small-pore systems, albeit with small binding energies. This further confirms that fluorine should be the dopant of choice for H-GNM systems.

Based on our results of the two studied pore sizes, we can infer how the doping occurs with larger ones. For the n-doping case, dopants will first bind to the pore edges. As the load is increased, clusters will form in the pore, giving rise to similar doping physics discussed in this work. For maximum loading, the number of dopants grows cubically with the cluster/pore radius, whereas the number of binding sites increases only linearly. As a result, the binding energy per dopant will decrease as the cluster size increases. On the other hand, the p-dopant, fluorine, will continue to bind to the pore edge with increasing pore radius. Fabrication of GNM pores is bound to produce pores with some level of disorder. As a result, a GNM will have a distribution of pore sizes and pore lattice constants, changing the symmetry requirement for the electronic gap formation. The effect of this is twofold. First, the gap will partially close, and second, the dopant load will vary for different pores. The latter effect will have negligible impact on potential applications of GNMs, as the resulting structures are still stably doped, whereas the former can seriously prevent the development of GNM-based transistors with technologically acceptable ON/OFF ratios. Chelation-doped crown GNMs may be considered for many other applications, such as transparent electrodes. On the other hand, their use in microelectronic applications requires controlled doping of the sheets. Thus, the pores, post fabrication, must be filled with the desired number of dopants, no more, no fewer. To achieve this result, it is convenient to select the binding energy dopant level with the lowest free energy such that sufficient thermal annealing, for instance, will allow the GNM/dopant system to relax to the correct number of dopants per pore. This approach avoids difficult stoichiometric control and other problems associated with achieving a doping state that is not the most thermodynamically stable. There are two caveats: first, choosing the right dopant whose lowest free energy state is a good fit for microelectronics, and second, the free energy of the next most stable dopant should be well separated from the lowest stable (large free energy gap) such that the thermal anneal will be effective. The results indicate that to achieve optimal control in transistor applications for large pores, dopants must be designed to fit in the pore of interest, allowing for carefully controlled device engineering.

1.3.2 Graphene/Nanotube Hybrid as TE Materials

Pristine undoped graphene has a relatively high sheet resistance, as the Fermi level is at the Dirac point where there are essentially no electronic states to serve as conducting channels. Doping, even by environmental effects, can easily move the Fermi level to a region where the resistance is considerably lower (due to an increase in the DOS). In practice, graphene grown by CVD has grain boundaries due to the use of polycrystalline substrates as the catalytic template material. It also has structural defects and various adsorbed species. The defects and grain boundaries significantly increase electron scattering in graphene, resulting in

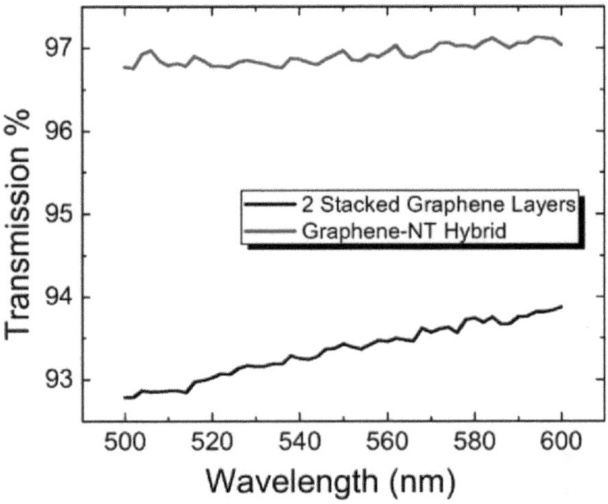

Figure 1.16 The optical density of the different films showing the increase in the sheet resistance of the graphene–CCNT hybrid film at a very high transparency. Reprinted with permission from Ref. [61], Copyright 2016, American Chemical Society.

its relatively high sheet resistance, even when doped [99]. A single-layer graphene prepared by CVD typically has a sheet resistance of 1 kΩ/sq [39]. If the adsorbed dopants are removed by annealing in vacuum, the sheet resistance can reach 5 kΩ/sq, which limits the utilization of graphene in many applications.

Several strategies have been developed to reduce graphene's sheet resistance. One is to stack graphene sheets to form a multilayer structure and thereby introduce more channels for charge transport [35, 38, 43]. Another is to dope the multilayers to shift the Fermi level and increase the DOS and thereby the conductivity [17, 43, 49, 57]. This approach has been successfully used to reduce sheet resistance to 100 Ω/sq, at 80% transparency [49, 92]. Patterning graphene with metal busbar structures via screen printing has also been found to be a successful strategy reducing sheet resistance to 22 Ω/sq, (limited by graphene/busbar contact resistance) while maintaining high transparency (90%) [40].

We began the characterization of our first hybrid sample by measuring its optical transparency. Figure 1.16 shows the transparency of the hybrid system as a function of wavelength over

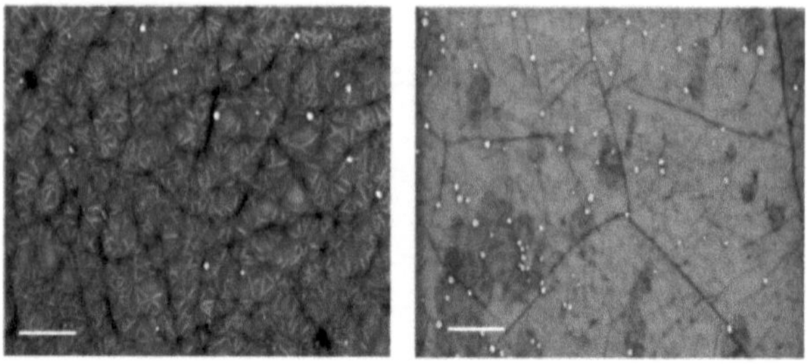

Figure 1.17 (a) SEM image of graphene–CCNT hybrid before annealing. (b) SEM image of graphene–CCNT hybrid after annealing where the nanotubes in the monolayer are not visible anymore due to the coupling with graphene. The scale bar is 600 nm. Reprinted with permission from Ref. [61], Copyright 2016, American Chemical Society.

the visible range and compares it to the transparency of a bilayer graphene sample. The transparency of the hybrid system is 97%, higher than that of a graphene bilayer.

Adsorbents between the conducting carbon nanotubes (CCNTs) and the graphene form an electrically insulating layer that decouples the CCNTs from the graphene. This explains our result that the pristine graphene, the bilayer graphene, and the graphene–CCNT hybrid samples have almost the same resistance. To overcome this problem, the samples were annealed at 600°C for 10 min in vacuum to remove any impurities and improve CCNT/graphene contact. The resistance of the graphene–CCNT hybrid system after treatment is about half that of the annealed graphene, while exhibiting nearly the same transmission. We posit that annealing at this high temperature in vacuum removes undesirable adsorbents and helps the nanotubes achieve good contact with the graphene. This was confirmed by the SEM images of the graphene–CCNT system before annealing (Fig. 1.17a). It is clear that CCNTs can be distinguished in the image due to charging. After annealing, the nanotubes cannot be distinguished.

To study the effect of the CCNT density on the sheet resistance of the hybrid material, we prepared samples with different CCNT monolayer densities. In Fig. 1.18a, we show our transmission and sheet resistance results for hybrid samples with different CCNT

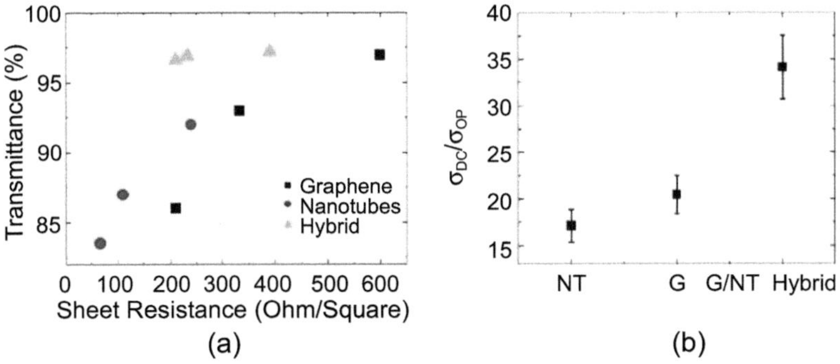

Figure 1.18 (a) Transmission versus sheet resistance. (b) The optical density of the different films showing the increase in the sheet resistance of the graphene–CCNT hybrid film at a very high transparency. Reprinted with permission from Ref. [61], Copyright 2016, American Chemical Society.

densities, multilayer graphene, and thick carbon CCNT films. All samples were first annealed and then doped by nitric acid as before. All hybrid samples have the same transparency as single-layer graphene, confirming their monolayer nature. The sheet resistance of the lowest-density CCNT hybrid sample is about half that of single-layer graphene. Increasing the monolayer density threefold brings the sheet resistance down to about one-third that of single-layer graphene. Further increase in the CCNT monolayer density has a minor effect on the sheet resistance of the hybrid. By comparison, similar resistance reduction is achieved with a thick CCNT film with transparency of 90%, and few layers graphene with a transparency of 85%.

We now explain our model. As a first approximation, and since the nanotubes are randomly dispersed on the graphene sheet, the coupling between the nanotubes and the underlying graphene is taken to be constant—a mean field type of approximation. Furthermore, all nanotubes are taken to be conducting, which is justified by the fact that the graphene–nanotube system is chemically doped, although purification methods are being developed to separate conducting from insulating tubes [89]. Since we are interested in the low nanotube-density limit where the nanotube- network percolation itself is absent or weak, contributions from the nanotubes to the graphene conductance will arise from the nanotubes crossing graphene scattering boundaries (e.g., a nanotube crossing over a grain boundary). That is, a nanotube creates another conductance path between neighboring regions, and thus one can think of the nanotube as a local modification of the resistance of the graphene sheet. We further assume that the contact resistance between crossing CCNTs is much higher than that between a CCNT and graphene [24, 62]. This is confirmed by the observed significant decrease in the hybrid resistance for CCNT film densities that are well below the CCNT-network percolation limit.

The graphene sheet is treated as a continuum resistive medium, which we discretize into a square grid of tiles. The tile edge, a_G, is taken to be smaller than the smallest length in the system (the nanotube length L_T). Each tile is connected to its four nearest neighbors using a resistance r_G (assuming that the tile itself has negligible internal resistance), determined by the experimentally measured neat graphene sheet resistance. The graphene–CCNT hybrid system is constructed by adding the nanotubes to the graphene sheet at random positions and with random orientations. We model this by adding local parallel resistance r_{GT} when a randomly placed tube crosses two graphene tiles. Nanotubes are taken to be of uniform size, $L_T \sim 350$ nm, as inferred from SEM images of the physical system. Figures 1.19a,b present a schematic of a graphene sheet with grains and the nanotubes acting as bridges connecting the grains.

The sheet resistance of the modeled hybrid system is determined using a 4-probe resistance calculation. Current is injected from the outer probes, and the voltage drop on the inner two probes is calculated using Kirchhoff's laws. The sheet resistance is thus

$$R = \frac{V_c - V_d}{I}$$

where I is the current entering node a and leaving node b. The reduction in the sheet resistance of the hybrid system can be described using the fractional decrease in the resistance caused by the CCNT film, defined as

$$\eta = \frac{R_G - R_H}{R_G}$$

where R_G is the sheet resistance of graphene, and R_H is the sheet resistance of the hybrid system.

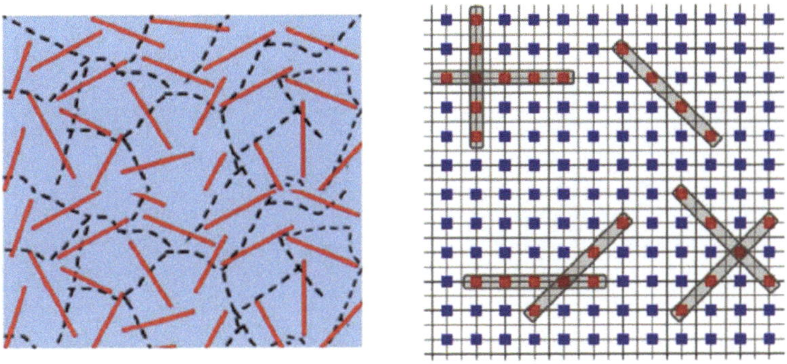

(a) (b)

Figure 1.19 (a) Schematic showing graphene sheet with a layer of nanotubes. The dotted lines mark the graphene grains and the CCNTs are shown as red lines. (b) A 2D resistance network model of the graphene sheet. The CCNTs increase the sheet conductance by locally changing the model resistance. Reprinted with permission from Ref. [61], Copyright 2016, American Chemical Society.

A sample size of side $L_S = 40$ microns was simulated. The tile size is taken to be $a_G = 0.25$ microns matching a typical graphene grain size determined from SEM images [31, 99]. The tile size is smaller than the nanotube length L_T, so that each CCNT is guaranteed to bridge at least two tiles, thereby simulating the coupled graphene–CCNT system. Reassuringly, further decrease in the tile size a_G does not lead to significant change in the calculated sheet resistance. The resistance parameter r_G is set by fitting the calculated neat graphene sheet resistance to the measured value. The CCNT density r_{2D} and length $L_T = 0.35$ microns are inferred from the SEM images of Fig. 1.17a. The inter-probe distance d_{4p} of the 4-probe calculation is taken to be 4 microns, so that it spans many graphene–CCNT grains. This way, we have $a_G < L_T \ll d_{4p} \ll L_S$. This order secures a quantitatively correct description of the system as experimentally realized and measured.

In Fig. 1.20, we plot the experimental values of h for the annealed and doped hybrid system with the results of our calculation. The resistance parameter r_{GT} describing the graphene–CCNT coupling is obtained by fitting the model to experiment as described above. For low CCNT densities, the model predicts a linear dependence

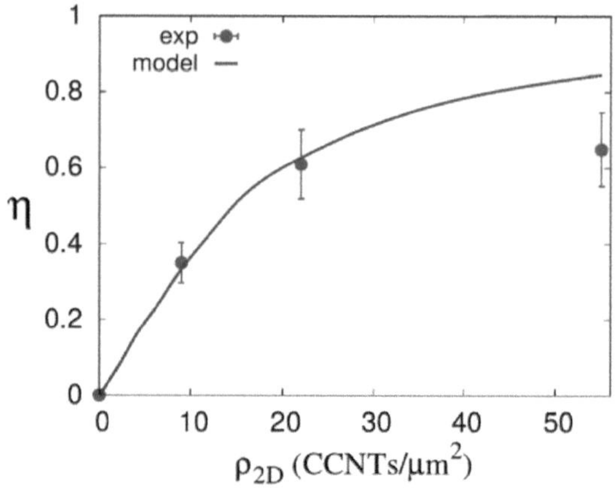

Figure 1.20 A plot of η versus the CCNT 2D density ρ. The model predicts a linear dependence on the CCNT density at lower densities below or at weak percolation of the CCNT network. For higher densities, not all CCNTs contribute to lowering the graphene resistance, η saturates. The deviation between the model and the experiment at high density is because 3D effects are not considered in the model. Reprinted with permission from Ref. [61], Copyright 2016, American Chemical Society.

of η on the density. As the density increases above the percolation threshold, not all CCNTs are in contact with the graphene, and since the CCNT–CCNT resistance is much higher than that of the CCNT–graphene, the fractional decrease in the resistance will reach a constant value, which is observed in our experimental data. Such 3D effects are not included in our simple model, which predicts a further increase in h at high CCNT densities, where some CCNTs lie on each other rather than on the graphene sheet. The inclusion of such effects would decrease the resistance gain obtained by the CCNT layer, but its treatment would be rather subjective in our phenomenological model. For example, it would have to involve how the CCNTs bend around each other, and how the bend is affected by details such as the number of underlying or overlying CCNTs, as well as the crossing angles. In addition, a model describing how this would affect the CCNT resistance would have to be constructed. However, these details do not change the overall picture captured by the current treatment, but explain the deviation of our model from our experimental data at high CCNT densities.

1.4 Graphene Structures as Templates for Catalytic Nanoparticles

Catalytic particles are widely used in various chemical processes. To increase the surface area of the catalyst, nano-sized particles are used. As the particle size gets smaller, its mobility on the supporting template increases, eventually leading to the agglomeration of the nanoparticles under their van der Waals interactions [64, 76, 93, 100]. This significantly reduces their effective surface area. Graphene has been considered a template for catalytic nanoparticles [3, 20]. The idea is to increase the binding between the catalytic nanoparticles and their supporting template, thus decreasing/preventing their agglomeration.

A GNM with nanometer-sized pores can be used to create a trap for charged nanoparticles [1], by using an electron beam, or by plasma etching. During the plasma etching of the graphene sheet to form GNM, free radical carbons are formed at the edge of the GNM pore. These free radicals are highly reactive, and depending on to what chemicals they are exposed, they can be converted to many functionalities. By selective passivation of these pores, the termination of the carbon dangling bonds can be controlled, leading, in some cases, to the doping of the graphene sheet. For example, generating the pores in air leads to the passivation of each pore with O, H, and OH groups. Charge is concentrated at the edge of the nanopores, through the dipoles formed by the difference in electronegativity between the edge carbons and the passivating species. These pores act as charged nests and can form "docking" locations for the targeted charged nanoparticles. The docking occurs using the electrostatic attraction between the charged nanoparticles and the oppositely charged nanopore.

In contrast to existing approaches, this approach includes structural control of the locations formed by the targeted nanoparticles. As the size and locations of the nanopores can be accurately chosen, the resulting final shape of the nanoparticles can be highly controlled.

As an example, Fig. 1.21 shows the unit cell of an NH_2-passivated GNM [1]. The GNM can be created in an ammonia atmosphere.

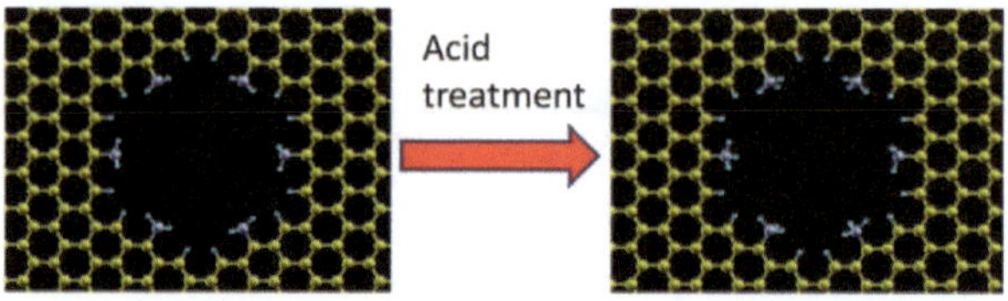

Figure 1.21 An NH_2-functionalized GNM. Acid treatment forms positively charged NH_3^+ groups at the pore edge, which can attract and anchor negatively charged catalytic nanoparticles.

As a result, carbon atoms along the edge of the nanopores are passivated by "H" and "NH_2" groups. Such a system can be referred to herein as NH_2-GNM. To create the positively charged trap, the NH_2-GNM is treated in an acid such as HX, where X is a halogen. This results in positively charged edges. We label this structure as $(NH_3\text{-}GNM)^+$. Negatively charged nanoparticles can now be anchored in the positively charged pores according to the reaction:

$$(NH_3 - GNM)^+X^- + Z^+Y^- \rightarrow (NH_3 - GNM)^+Y^- + Z^+X^-,$$

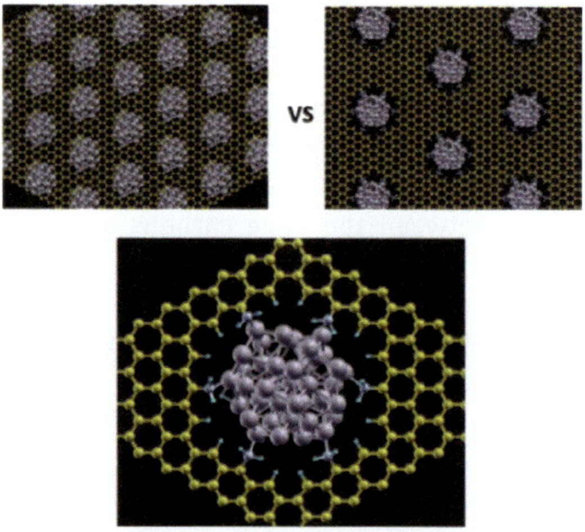

Figure 1.22 Top: a 1 nm Pt nanoparticles on pristine graphene and on an NH_2-functionalized GNM. The binding energy per bond in the GNM case is double that in the case of graphene. Bottom: the unit cell of the Pt@NH_2-GNM system.

where Z^+Y^- represents a weak base solution containing the nanoparticles Y (for example, a sodium citrate with gold nanoparticles).

We now compare two templates—pristine graphene and the NH_2-GNM—considered supports for a 1 nm Pt particle. The first-principle calculations show that some of the H_2 is released, allowing surface Pt atoms in the nanoparticle to covalently bond with the pore edge nitrogen atoms, with a binding energy of 4 eV per formed bond (six bonds are formed). The corresponding value on graphene is only half of that, indicating that the NH_2-functionalized GNM has a great potential for strongly anchoring catalytic nanoparticles.

The main advantage of this approach [1] is to provide structural stability in the formed nanolayer. The strong binding between the nanoparticles and the nanopore functional groups guarantees a certain level of stability. Such an interaction can be tailored to best suit the nanoparticles of interest via a customized choice of the functional groups passivating the pore edges. Therefore, positively and negatively charged graphene nanomesh nests can be formed. In applications where different nanoparticle species are required, a hybrid structure can be formed.

References

1. Afzali-Ardakani, A. et al. (2014). Controlled assembly of charged nanoparticles using functionalized graphene nanomesh, US patent 8,835,686.
2. Allen, B. L. et al. (2007). Carbon nanotube field-effect-transistor-based biosensors, *Adv. Mater.*, **19**, pp. 1439–1451.
3. Antolini, E. (2012). Graphene as a new carbon support for low-temperature fuel cell catalysts, *Appl. Catal. B Environ.*, 123–124, pp. 52–68.
4. Barnes, W. L. (1998). Fluorescence near interfaces: The role of photonic mode density, *J. Mod. Opt.*, **45**, pp. 661–699.
5. Barnes, W. L. et al. (2003). Surface plasmon subwavelength optics, *Nature*, **424**, pp. 824–830.
6. Bekyarova, E. et al. (2009). Chemical modification of epitaxial graphene: Spontaneous grafting of aryl groups, *Am. Chem. Soc.*, **131**, pp. 1336–1337.

7. Berg, J. M. et al. (2002). Biochemistry, 5th ed., New York: W H Freeman.
8. Chen, Z. et al. (2010). Energy transfer from individual semiconductor nanocrystals to graphene, *ACS Nano*, **4**, pp. 2964–2968.
9. Chhabra, R. et al. (2009). Distance-dependent interactions between gold nanoparticles and fluorescent molecules with DNA as tunable spacers, *Nanotechnology*, **20**, pp. 485201–485210.
10. Copel, M. (2000). Medium-energy ion scattering for analysis of microelectronic materials, *IBM J. Res. Dev.*, **44**, pp. 571–528.
11. Copel, M. et al. (2011). Medium energy ion scattering of Gr on SiC(0001) and Si(100), *Appl. Phys. Lett.*, **98**, pp. 113103–113105.
12. Cui, Y. et al. (2003). High performance silicon nanowire field effect transistors, *Nano Lett.*, **3**, pp. 149–152.
13. Dostalek, J. and Knoll, W. (2008). Biosensors based on surface plasmon-enhanced fluorescence spectroscopy, *Biointerphases*, **3**, pp. 12–22.
14. Dostalek, J. et al. (2007). Long range surface plasmons for observation of biomolecular binding events at metallic surfaces, *Plasmonics*, **2**, pp. 97–106.
15. Duan, X. et al. (2012). Quantification of the affinities and kinetics of protein interactions using silicon nanowire biosensors, *Nat. Nanotech.*, **7**, pp. 401–407.
16. Sheka, E. F. (2014). The uniqueness of physical and chemical natures of graphene: Their coherence and conflicts, *Int. J. Quantum Chem.*, **114**, pp. 1079–1095.
17. Eberlein, T. et al. (2008). Doping of graphene: Density functional calculations of charge transfer between GaAs and carbon nanostructures, *Phys. Rev. B*, **78**(4) , pp. 045403.
18. Eldeeb, M. S. et al. (2018). *Carbon*, **133**, pp. 369–378.
19. Withers, F. et al. (2015). Light-emitting diodes by band-structure engineering in van der Waals heterostructures, *Nat. Mater.*, **14**(3), pp. 301–306.
20. Fan, X. (2015). Multiple roles of graphene in heterogeneous catalysis, *Chem. Soc. Rev.*, **44**, pp. 3023–3035.
21. Farmer, D. et al. (2009). Chemical doping and electron–hole conduction asymmetry in graphene devices, *Nano Lett.*, **9**, pp. 388–392.
22. Farmer, D. et al. (2009). Behavior of a chemically doped graphene junction, *Appl. Phys. Lett.*, **94**, pp. 213106.

23. Folkers, J. P. et al. (1995). Self-assembled monolayers of long-chain hydroxamic acids on the native oxides of metals, *Langmuir*, **11**, pp. 813–824.
24. Fuhrer, M. S. et al. (2000). Crossed nanotube junctions, *Science*, **288**, pp. 494–497.
25. Geim, A. K. (2009). Graphene: Status and prospects, *Science*, **324**, pp. 1530–1534.
26. Geim, A. K. and Novoselov, K. S. (2007). The rise of graphene, *Nat. Mater.*, **6**, pp. 183–191.
27. Homola, J. (2003). Present and future of surface plasmon resonance biosensors, *Anal Bioanal Chem.*, **377**, pp. 528–539.
28. Homola, J. et al. (1999). Surface plasmon resonance sensors: Review, *Sens. Actuators B*, **54**, pp. 3–15.
29. Honga, H. et al. (2009). Surface modification of the polyethyleneimine layer on silicone oxide film via UV radiation, *Appl. Surf. Sci.*, **255**, pp. 6103–6016.
30. Horowitz, G. (1999). Organic field-effect transistors, *Adv. Mater.*, **10**, pp. 365–377.
31. Huang, P. Y. et al. (2011). Grains and grain boundaries in single-layer graphene atomic patchwork quilts, *Nature*, **469**, pp. 389–392.
32. Huang, Y. et al. (2010). Nanoelectronic biosensors based on CVD grown graphene, *Nanoscale*, **2**, pp. 1485–1488.
33. Wassei, J. K. and Kaner, R. B. (2010). Graphene, a promising transparent conductor, *Mater. Today*, **13**(3), pp. 52–59.
34. Javey, A. et al. (2003). Ballistic carbon nanotube field-effect transistors, *Nature*, **424**, pp. 654–657.
35. Jung, M. et al. (2009). Charge transfer chemical doping of few layer graphenes: Charge distribution and band gap formation, *Nano Lett.*, **9**(12), pp. 4133–4137.
36. Justino, C. L. (2017). Graphene based sensors and biosensors, *TrAC Trends Anal. Chem.*, **91**, pp. 53–66.
37. Kasry, A. and Knoll, W. (2006). Long range surface plasmon fluorescence spectroscopy, *Appl. Phys. Lett.*, **89**, pp. 101106–8.
38. Kasry, A. et al. (2010). Chemical doping of large-area stacked graphene films for use as transparent, conducting electrodes, *ACS Nano*, **4**, pp. 3839–3384.
39. Kasry, A. et al. (2011). Detection of biomolecules via benign surface modification of graphene, *Chem. Mater.*, **23**(22), pp. 4879–4881.

40. Kasry, A. et al. (2012). High performance metal microstructure for carbon-based transparent conducting electrodes, *Thin Solid Films*, **520**(15), pp. 4827–4830.
41. Kasry, A. et al. (2012). Highly efficient fluorescence quenching with graphene, *J. Phys. Chem. C*, **116**, pp. 2858–2862.
42. Khan, H. et al. (2010). In situ, label-free DNA detection using organic transistor sensors, *Adv. Mater.*, **22**, pp. 4452–4456.
43. Kim, K. K. et al. (2010). Enhancing the conductivity of transparent graphene films via doping, *Nanotechnology*, **21**(28), pp. 285205.
44. Kim, K. S. et al. (2009). Large-scale pattern growth of graphene films for stretchable transparent electrodes, *Nature*, **457**, pp. 706–710.
45. Knoll, W. (1998). Interfaces and thin films as seen by bound electromagnetic waves, *Ann. Rev. Phys. Chem.*, **49**, pp. 569–638.
46. Knoll, W. et al. (2008). New concepts with surface plasmons and nano-biointerfaces, *J. Nonlinear Opt. Phys. Mater.*, **17**, pp. 121–129.
47. Kovacs, G. J. and Scott, G. D. (1977). Optical excitation of surface plasma waves in layered media, *Phys. Rev. B*, **16**(4), pp. 1297–1311.
48. Lakowicz, J. R. (1999). *Principles of Fluorescence Spectroscopy*, 2nd ed., New York: Kluwer.
49. Late, D. J. et al. (2011). Molecular charge-transfer interaction with single-layer graphene, *J. Exp. Nanosci.*, **6**(6), pp. 641–651.
50. Levine, P. M. et al. (2009). Real-time, multiplexed electrochemical DNA detection using an active complementary metal-oxide-semiconductor biosensor array with integrated sensor electronics, *Biosens Bioelectron.*, **24**, pp. 1995–2001.
51. Li, J. (2005). *Basic Molecular Dynamics. Handbook of Materials Modeling*, Springer. 565588.
52. Li, X. et al. (2009). Transfer of large-area graphene films for high-performance transparent conductive electrodes, *Nano Lett.*, **9**, pp. 4359–4363.
53. Li, X. S. et al. (2009). Large-area synthesis of high-quality and uniform graphene films on copper foils, *Science*, **324**, pp. 1312–1314.
54. Liebermann, T. and Knoll, W. (2000). Surface-plasmon field-enhanced fluorescence spectroscopy, *Colloid. Surf. A*, **171**, pp. 115–130.
55. Liebermann, T. et al. (2000). Complement hybridization from solution to surface-attached probe-oligonucleotides observed by surface-plasmon-field-enhanced fluorescence spectroscopy, *Colloid. Surf. A*, **169**, pp. 337–350.

56. Lodish, H. et al. (1999). *Molecular Cell Biology*, 4th ed., Palgrave Macmillan.
57. Lu, Y. H. et al. (2009). Tuning the electronic structure of graphene by an organic molecule, *J. Phys. Chem. B*, **113**(1), pp. 2–5.
58. Liu, M. et al. (2014). First-principles studies of Li nucleation on graphene, *J. Phys. Chem. Lett.*, **5**(7), pp. 1225–1229.
59. Matyusa, L. et al. (2006). Steady-state fluorescence quenching applications for studying protein structure and dynamics, *J. Photochem. Photobiol. B*, **83**, pp. 223–236.
60. Maarouf, A. A. et al. (2013). Crown graphene nanomeshes: Highly stable chelation-doped semiconducting materials, *J. Chem. Theory Comput.*, **9**, pp. 2398–2403.
61. Maarouf, A. A. et al. (2016). A graphene-carbon nanotube hybrid material for photovoltaic applications, *Carbon*, **102**, pp. 74–80.
62. Maarouf, A. A. et al. (2011). Low-energy coherent transport in metallic carbon nanotube junctions, *Phys. Rev. B*, **83**, pp. 045402.
63. Machado, B. F. and Serp, P. (2012). Graphene-based materials for catalysis, *Catal. Sci. Technol.*, **2**, pp. 54–75.
64. Maillard, F. et al. (2005). Influence of particle agglomeration on the catalytic activity of carbon-supported Pt nanoparticles in CO monolayer oxidation, *Phys. Chem. Chem. Phys.*, **7**, pp. 385–393.
65. Martel, R. et al. (1998). Single- and multi-wall carbon nanotube field-effect transistors, *Appl. Phys. Lett.*, **73**, pp. 2447–2449.
66. Mu, J. et al. (2016). An elastic transparent conductor based on hierarchically wrinkled reduced graphene oxide for artificial muscles and sensors, *Adv. Mater.*, **28**(43), pp. 9491–9497.
67. Neto, A. et al. (2009). The electronic properties of graphene, *Rev. Mod. Phys.*, **81**, pp. 109–162.
68. Neumann, T. et al. (2002). Surface–plasmon fluorescence spectroscopy, *Adv. Mat.*, **12**, pp. 575–586.
69. Nistor, R. A. et al. (2011). The role of chemistry in graphene doping for carbon-based electronics, *J. ACS Nano*, **5**, pp. 3096–3103.
70. Novoselov, K. S. et al. (2004). Electric field effect in atomically thin carbon films, *Science*, **306**, pp. 666–669.
71. Ohno, Y. et al. (2009). Electrolyte-gated graphene field-effect transistors for detecting pH and protein adsorption, *Nano Lett.*, **9**, pp. 3318–3322.

72. Ohno, Y. et al. (2010). Chemical and biological sensing applications based on graphene field-effect transistors, *Biosens. Bioelectron.*, **26**, pp. 1727–1730.

73. Otto, A. Z. (1968). Excitation of nonradiative surface plasma waves in silver by the method of frustrated total reflection, *Physik*, **216**, pp. 398–410.

74. Pang, S. et al. (2009). Patterned graphene electrodes from solution-processed graphite oxide films for organic field-effect transistors, *Adv. Mater.*, **21**, pp. 3488–3491.

75. Pantha, N. et al. (2015). First-principles study of the interaction of hydrogen molecular on Na-adsorbed graphene, *Appl. Nanosci.*, **5**(4), pp. 393–402.

76. Pellegrino, F. et al. (2017). Influence of agglomeration and aggregation on the photocatalytic activity of TiO_2 nanoparticles, *Appl. Catal. B Environ.*, **216**, pp. 80–87.

77. Persson, B. N. J. and Lang, N. D. (1982). Electron-hole-pair quenching of excited states near a metal, *Phys. Rev. B*, **26**, pp. 5409–5415.

78. Quintana, M. et al. (2010). Functionalization of graphene via 1,3-dipolar cycloaddition, *ACS Nano*, **4**, pp. 3527–3533.

79. Rather, H. (1988). *Surface Plasmons*, Springer-Verlag Berlin Heidelberg.

80. Ritchie, R. H. (1957). Plasma losses by fast electrons in thin films, *Phys. Rev.*, **106**, pp. 874.

81. Chae, S. H. and Lee, Y. H. (2014). Carbon nanotubes and graphene towards soft electronics, *Nano Convergence*, **1**(1), pp. 15.

82. Shao, Y. et al. (2009). Graphene based electrochemical sensors and biosensors: A review, *Electroanalysis*, **22**, pp. 1027–1036.

83. Sinitskii, A. et al. (2010). Kinetics of diazonium functionalization of chemically converted graphene nanoribbons, *ACS Nano*, **4**, pp. 1949–1954.

84. Stengel, G. and Knoll, W. (2005). Surface plasmon field-enhanced fluorescence spectroscopy studies of primer extension reactions, *Nucleic Acids Res.*, **33**, pp. e69–e69.

85. Sudibya, H. G. et al. (2011). Electrical detection of metal ions using field-effect transistors based on micropatterned reduced graphene oxide films, *ACS Nano*, **5**, pp. 1990–1194.

86. Suvarnaphaet, P. and Pechprasarn, S. (2017). Graphene-based materials for biosensors: A review, *Sensors*, **17**, pp. 2161–2185.

87. Swathi, R. S. and Sebastian, K. L. (2008). Resonance energy transfer from a dye molecule to graphene, *J. Chem. Phys.*, **129**, pp. 054703-1–054703-9.

88. Swathi, R. S. and Sebastian, K. L. (2009). Long range resonance energy transfer from a dye molecule to graphene has (distance)(-4) dependence, *J. Chem. Phys.*, **130**, pp. 086101-3.

89. Tulevski, G. S. et al. (2013). High purity isolation and quantification of semiconducting carbon nanotubes via column chromatography, *ACS Nano*, **7**(4), pp. 2971–2976.

90. Vareiro, M. et al. (2008). Surface plasmon resonance enhanced fluorescence implementation of a single-step competition assay: Demonstration of fatty acid measurement using an anti-fatty acid monoclonal antibody and a Cy5-labeled fatty acid, *Anal. Biochem.*, **377**(2), pp. 243–250.

91. Vasilev, K. et al. (2004). Fluorescence intensities of chromophores in front of a thin metal film, *J. Chem. Phys.*, **20**, pp. 3439–3445.

92. Voggu, R. et al. (2008). Effects of charge transfer interaction of graphene with electron donor and acceptor molecules examined using Raman spectroscopy and cognate techniques, *J. Phys. Condens. Matter.*, **20**, pp. 472204-9.

93. Wang, et al. (2016). Agglomeration mechanism of nanoparticles by adding coarse fluid catalytic cracking particles, *Chem. Eng. Technol.*, **39**, pp. 1490–1496.

94. Wehling, T. O. et al. (2010). Resonant scattering by realistic impurities in graphene, *Phys. Rev. Lett.*, **105**, pp. 056802-5.

95. Weitz, D. A. et al. (1982). Fluorescent lifetimes of molecules on silver-island films, *Opt. Lett.*, **7**, pp. 89–91.

96. Fan, X. et al. (2015). Multiple roles of graphene in heterogeneous catalysis, *Chem. Soc. Rev.*, **44**(10), pp. 3023–3035.

97. Yang, W. et al. (2010). Carbon nanomaterials in biosensors: Should you use nanotubes or graphene? *Angew. Chem. Int. Ed.*, **49**, pp. 2114–2138.

98. Yu, F. et al. (2004). Attomolar sensitivity in bioassays based on surface plasmon fluorescence spectroscopy, *J. Am. Chem. Soc.*, **126**, pp. 8902–8903.

99. Yu, Q. et al. (2011). Control and characterization of individual grains and grain boundaries in graphene grown by chemical vapour deposition, *Nat. Mater.*, **10**(6), pp. 443–449.

100. Zhang, C. et al. (2012). Graphene supported nanoparticles of Pt–Ni for CO oxidation, *Appl. Surf. Sci.*, **258**, pp. 7795–7800.

101. Zhang, Y. et al. (2005). Experimental observation of the quantum Hall effect and Berry's phase in graphene, *Nature*, **438**, pp. 201–204.

102. Zhang, Z. et al. (2003). Surface plasmon resonance studies of protein binding on plasma polymerized di(ethylene glycol) monovinyl ether films, *Langmuir*, **19**, pp. 4765–4470.

103. Zhuang, X. et al. (2000). Fluorescence quenching: A tool for single-molecule protein-folding study, *Proc. Natl. Acad. Sci. U.S.A*, **9**, pp. 14241–14244.

Chapter 2

Essential Properties of Fluorinated Graphene and Graphene Nanoribbons

Duy Khanh Nguyen,[a] Ngoc Thanh Thuy Tran,[a] Thanh Tien Nguyen,[b] Yu-Huang Chiu,[c] and Ming-Fa Lin[a]

[a] *Department of Physics, National Cheng Kung University, Tainan, Taiwan*
[b] *Department of Physics, College of Natural Sciences, Can Tho University, Can Tho City, Vietnam*
[c] *Department of Applied Physics, National Pingtung University, Pingtung, Taiwan*
nguyenkhanhphysics2015@gmail.com

A systematic study is conducted on the fluorination-enriched essential properties of two-dimensional (2D) graphene and one-dimensional (1D) graphene nanoribbons using the first-principles method. The combined effects, which arise from the significant chemical bonds in C–C, F–C, and F–F bonds, the finite-size quantum confinement, and the edge structure, can greatly diversify geometric structures, electronic properties, and magnetic configurations. By the detailed analyses, the critical orbital hybridizations in determining the essential properties are accurately identified from the atom-dominated energy bands, the spatial charge distributions, and the orbital-projected density of states. The top-site F–C bonds, with the multi-orbital hybridizations, create the non-uniform buckled

2D and 3D Graphene Nanocomposites: Fundamentals, Design, and Devices
Edited by Olga E. Glukhova

ISBN 978-981-4800-41-9 (Hardcover), 978-0-429-20150-9 (eBook)
www.jennystanford.com

honeycomb lattice. There exist the C-, F-, and (C, F)-dominated energy bands. Fluorinated graphene belongs either to the p-type metals (with/without the ferromagnetic spin arrangement) or to the large-gap semiconductors (without magnetism), depending on the concentration and distribution of adatoms. Specially, fluorinated graphene nanoribbons, with armchair/zigzag edge, present five kinds of spin-dependent properties, covering the non-magnetic and ferromagnetic metals, non-magnetic semiconductors, and anti-ferromagnetic semiconductors with/without the spin splitting. The various band-edge states in 2D and 1D systems appear as the rich and unique structures in density of states. Part of theoretical predictions is consistent with the experimental measurements, and the others are worthy of further examinations. Also, the fluorination-created diverse properties clearly indicate the high potentials in various applications that will be discussed in detail, e.g., electronic and spintronic nanodevices.

2.1 Introduction

Since graphene was successfully produced, numerous studies, both theoretical [1] and experimental [2], have been carried out to modulate its properties, especially the electronic structures. Adatom doping on graphene sheets with different types of elements such as oxygen, hydrogen, alumina, or alkali is one of the efficient methods for diversifying the band structure of graphene [3]. Fluorinated graphenes are commonly synthesized using a two-step process: fluorination to obtain fluorinated graphite [4] and then exfoliation to achieve the monolayer one [5]. Monolayer F-doped graphene can also be directly synthesized by heating a mixture of graphene sheet and XeF_2 [6].

Graphene nanoribbons (GNRs) [7] have emerged as the ideal nanomaterials for removing the unexpected gapless feature of 2D graphene due to the complex relations of 1D honeycomb lattices, one-atom thicknesses, finite-size quantum confinements, and edge structures. GNRs are successfully fabricated by various experimental methods, including both top-down and bottom-up schemes [8].

From the geometric side, one of the simplest and intuitive methods for creating GNRs is the cutting of 2D graphene sheet that covers lithographic patterning [9], graphene etching [10], sonochemical breaking [11], and metal-catalyzed treatment [12]. Also the massive amount of GNRs can be produced by the chemical vapor deposition or unzipping of the multi-walled carbon nanotubes [13]. The essential properties of GNR materials can lead to many potential applications in nano-electronic [14] and spintronic [15] devices and gas sensor [16]. Furthermore, the special electronic properties of GNRs can be diversified by many different paths such as mechanical strain [17, 18], layer number [19, 20], curved surface [21], edge passivation [22, 23], stacking configuration [24], electric [25, 26] and magnetic [27, 28] fields, and adatoms adsorption [29, 30]. Chemical adsorption on the planar surface of GNRs is one of the effective routes that can result in the feature-rich properties. There are a lot of previous studies in terms of chemical-doped GNRs. For example, the transition metal adatom-doped GNRs induce the spin-splitting metallic 1D energy bands with free electron density [31]. Alkali-adsorbed GNRs in armchair and zigzag edges, respectively, exhibit the non-magnetic and ferromagnetic metal with high electron density [32]. The 1D semiconducting or metallic band structures are revealed in the ligand-protected aluminum-clusters-adsorbed AGNRs [33].

Interestingly, fluorine has the strongest electron affinity among other halogen or atoms; it is, thus, expected to form significant chemical bonds with carbon atoms and greatly diversify essential properties. The feature-rich electronic properties of 2D graphene and 1D GNRs are investigated in detail for various F adatom concentrations and distributions by the first-principles calculations. The geometric structures strongly depend on the top-site F–C bonds and so do the electronic properties. The atom-dominated energy bands, the orbital-projected DOS, and the spatial charge distributions can be used to comprehend the critical multi-orbital hybridizations in C–C, F–C, and F–F bonds, which are responsible for the enriched electronic and magnetic properties. The up-to-date theoretical and experimental studies as well as the potential applications are also discussed.

2.2 Computational Methods

Modeling simulation on computer schemes, providing an important prediction for further experimental studies, has been emphasized as a pioneering method in physics, chemistry, and materials sciences in the recent years. A specific system can be fully described by the quantum mechanic Schrödinger equation. However, such equation is impossible to solve for many-body particle systems. Hence, possible approximations are utilized to deal with the problem. Up to now, the first-principles calculations have been considered an ideal method for materials simulation since the complicated electron–electron and electron–ion interactions are thoroughly characterized by a series of approximations and simplifications. Furthermore, the Vienna ab initio simulation package (VASP) [34] computes an approximate solution to the many-body Schrödinger equation, either within the density functional theory (DFT) to solve the Kohn–Sham equations, or within the Hartree–Fock approximation to solve the Roothaan equations. In this study, the adatom-enriched essential properties are investigated by the first-principles DFT and VASP calculations. The exchange-correlation energy due to the electron–electron interactions is calculated from the Perdew–Burke–Ernzerhof functional under the generalized gradient approximation [35]. The projector-augmented wave pseudopotentials are used to evaluate the electron–ion interactions [36]. The wave functions are built from the plane wave with a maximum energy cutoff of 400 eV, in which they are tested to achieve a very high accuracy. All the atoms in the relaxation process are subjected to the adjustments of their positions in order to have an optimized structure with the lowest total ground-state energy. And then the most stable configuration is used as the input parameters for further calculations on the fundamental physical properties. The interactions between two neighboring images are eliminated by applying the vacuum distance along the z = 15 Å and (z = 15 Å, y = 15 Å) axes for 2D graphene and 1D graphene nanoribbons, respectively. To fully understand the surface adsorption effects on magnetism, the spin configurations are taken into account for the fluorine-adatoms-adsorbed graphene and GNRs. The first

Brillouin zone is sampled in a Gamma scheme along the 2D and 1D periodic directions by $12 \times 12 \times 1$ and $12 \times 1 \times 1$ k-points for structure relaxation, and then $100 \times 100 \times 1$ and $100 \times 1 \times 1$ for accurate evaluations on electronic properties, respectively. Besides, the convergence criterion for energy is set at 10^{-5} eV between two simulation steps, and the maximum Hellmann–Feynman forces acting on each atom are smaller than 0.01 eV/Å during the ionic relaxations.

2.3 Results and Discussions

2.3.1 Fluorinated 2D Graphene

The stable adsorption position of F atoms on graphene is determined by comparing the binding energies, in which the lower value corresponds to higher stability. The binding energy (E_b) of n adsorbed F adatoms is expressed as $E_b = (E_{tot}–E_{gra}–nE_{Fluo})/n$, where E_{tot}, E_{gra}, and E_{Fluo} are the total ground-state energies of the fluorinated graphene, pristine graphene, and isolated F adatom, respectively. Among the bridge, hollow, and top sites, our calculations show that the top site is the most stable one, in agreement with previous studies [37]. After self-consistent calculations, we found that all the fluorinated graphene systems, with various concentrations and distributions, have stable buckled structures as demonstrated in Fig. 2.1. The adatom distributions are divided into two main kinds: In one, F atoms are absorbed at the same sublattice (Figs. 2.1a, 2.1c–e); in another, adsorptions take place at two different sublattices simultaneously (Figs. 2.1b,f). The latter is predicted to be more stable than the former because of the lower binding energy, e.g., the 25% cases (Table 2.1). The nearest C–C bond lengths in graphene are expanded in the ranges of 1.45–1.58 Å with the increase in concentrations of F. The F–C bond lengths are shorter in the ranges of 1.38–1.57 Å with an increase in F concentration, indicating stronger F–C bonds at high F concentrations.

The 2D band structures along the high symmetry points are useful in examining the electronic properties. Monolayer graphene

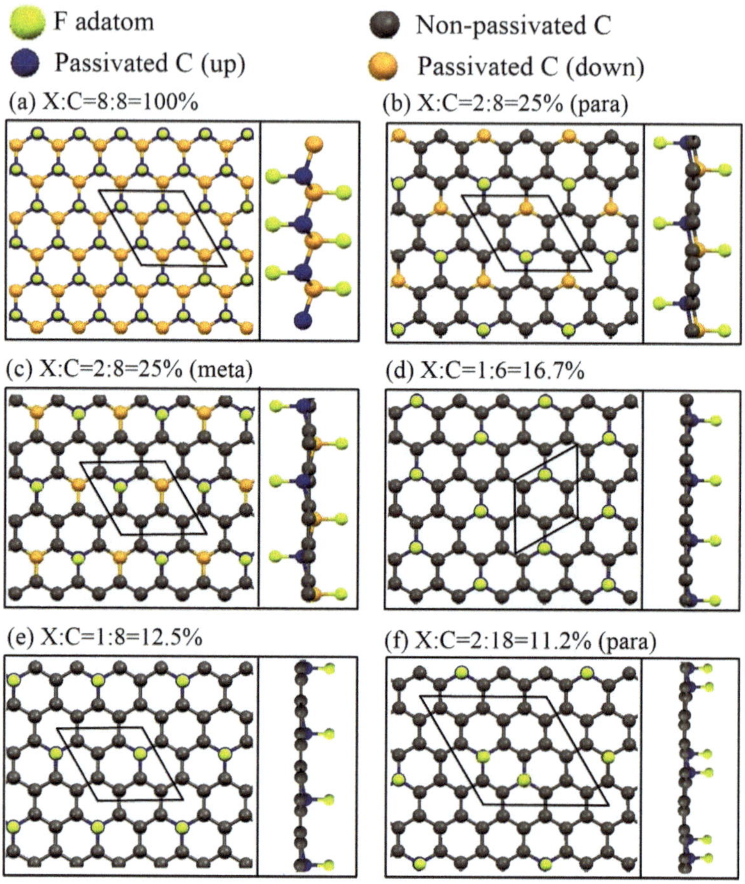

Figure 2.1 Geometric structures with top and side views for various concentrations and distributions: (a) F:C = 8:8 = 100% (double-side), (b) F:C = 2:8 = 25% (para-site), (c) F:C = 2:8 = 25% (meta-site), (d) F:C = 1:6 = 16.7%, (e) F:C = 1:8 = 12.5%, and (f) F:C = 2:18 = 11.2% (para-site).

has a Dirac-cone structure at the K point (Fig. 2.2a), owing to the extended π bondings of $2p_z$ orbitals in a hexagonal sheet. The linear energy bands turn into the parabolic dispersions in the increase in state energy, e.g., the parabolic bands near the *M* point. Furthermore, the σ bands, which arise from the sp^2 bondings, are initiated from the Γ point at the deeper energy. When F atoms are adsorbed on graphene, the energy dispersion relations exhibit dramatic changes

Table 2.1 Calculated C–C and C–F bond lengths, heights (shift on z-axis) of passivated carbons, binding energies, energy gaps, and total magnetic moments per unit cell of fluorinated graphene systems

				Bond length		Height	E_b	E_g	M_{tot}
Adatom	**F:C**	**%**	**C-F (Å)**	**1st C-C (Å)**	**2nd C-C (Å)**	**(Å)**	**(eV)**	**(eV)**	**(μ_B)**
F	8:8*	100	1.38	1.58		0.48	−3.00	3.12	0
	4:8* (para)	50	1.48	1.51		0.38	−2.71	2.77	0
	4:8 (meta)	50	1.48	1.5		0.28	−1.16	0	1.94
	2:8* (para)	25	1.48	1.52	1.41	0.36	−2.96	2.72	0
	2:8* (meta)	25	1.54	1.48	1.43	0.34	−2.68	0	0
	1:6	16.7	1.53	1.48	1.42	0.32	−2.45	0	0.22
	1:8	12.5	1.53	1.49	1.42	0.32	−2.64	0	0.5
	2:18* (para)	11.2	1.48	1.51	1.42	0.34	−2.46	0	0

Note: The double-side structures are labeled with *.

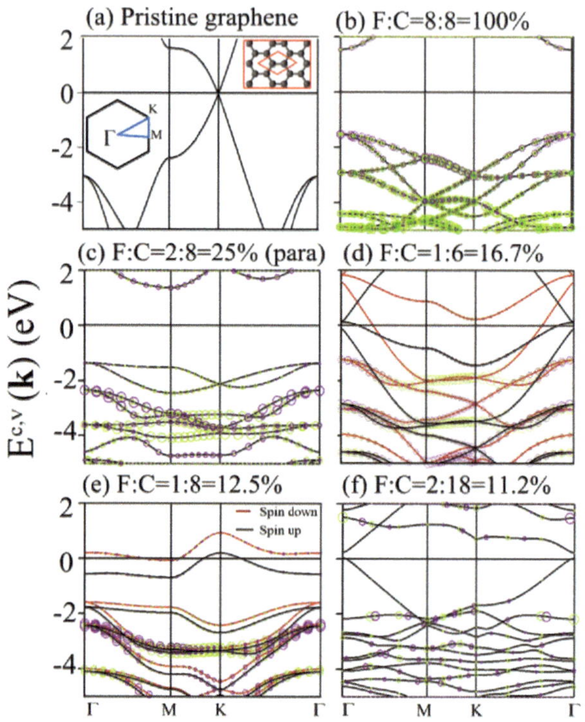

Figure 2.2 Band structures of various concentrations: (a) pristine graphene, (b) F:C = 8:8 = 100% (double-side), (c) F:C = 2:8 = 25% (para-site), (d) F:C = 1:6 = 16.7%, (e) F:C = 1:8 = 12.5%, (f) F:C = 2:18 = 11.2% (para-site). The green and purple circles correspond to the contributions of F and passivated C atoms, respectively.

with the variations in F-concentrations and F-distributions. The linear isotropic Dirac cone of graphene is destroyed at high F-concentration, mainly owing to the strong F–C bonds (discussed later). Under the full fluorination (Fig. 2.2b), a direct energy gap of 3.12 eV appears at the Γ point, in agreement with the previous theoretical calculations [38] and experimental measurements [39]. With the decrease in fluorination, the semiconducting or metallic behaviors are mainly determined by the C-dominated energy bands (Figs. 2.2c–f), since the π-electronic structure (the $2p_z$-orbital bonding) is gradually recovered. The F–F bonding quickly declines, while the strength of the F–C bond is almost the same, as suggested

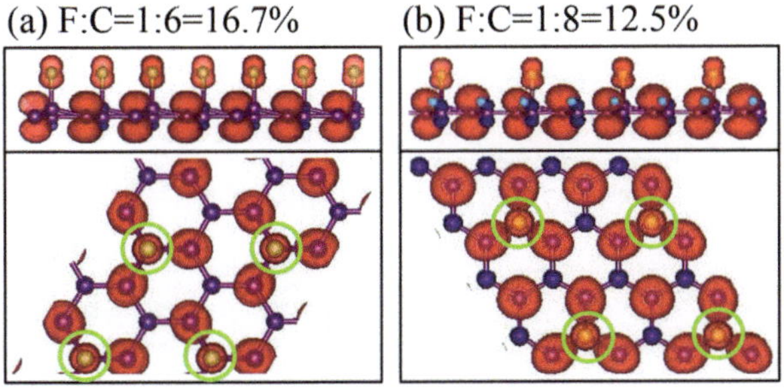

Figure 2.3 The spin-density distributions with top and side views for (a) F:C = 16.7% and (b) F:C = 12.5%. The red isosurfaces represent the charge density of spin-up configuration.

by the weakly dispersive (F,C)-co-dominated bands. The fluorinated systems, as shown in Figs. 2.2d,e, have the spin-split energy bands across the Fermi level simultaneously; that is, they present the metallic ferromagnetism. Such bands are dominated by carbon atoms ($2p_z$ orbitals), but not F adatoms. The occupied/unoccupied carrier densities in the spin-up and spin-down bands differ from each other (the black and red curves in Figs. 2.2d,e). There exist ferromagnetic spin configurations under the same sublattice adsorption, as observed in Figs. 2.3a,b for 16.7% and 12.5% F concentration, respectively.

The charge density (Figs. 2.4a–c) and the charge density difference (Figs. 2.4d,e) can provide very useful information about the chemical bondings. The latter is obtained by subtracting the charge density of pristine graphene and F adatoms from that of fluorinated graphene system. As F atoms are adsorbed on graphene, their orbitals have strong hybridizations with those of passivated C, as seen from the red region enclosed by the dashed black rectangles (Figs. 2.4b,c). Compared with pristine graphene (Fig. 2.4a), such strong F–C bonds lead to the deformed π bonds and thus the serious distortion of the Dirac-cone structures (Fig. 2.2a). Between two non-passivated C atoms, ρ shows a strong σ bonding (black rectangles), which is slightly reduced after the formation of F–C bonds. There also exist the significant F–F bonds (purple rectangles) where the adatoms are sufficiently close to each other. These can create the (F,C)-co-dominated energy bands, the F-induced energy bands,

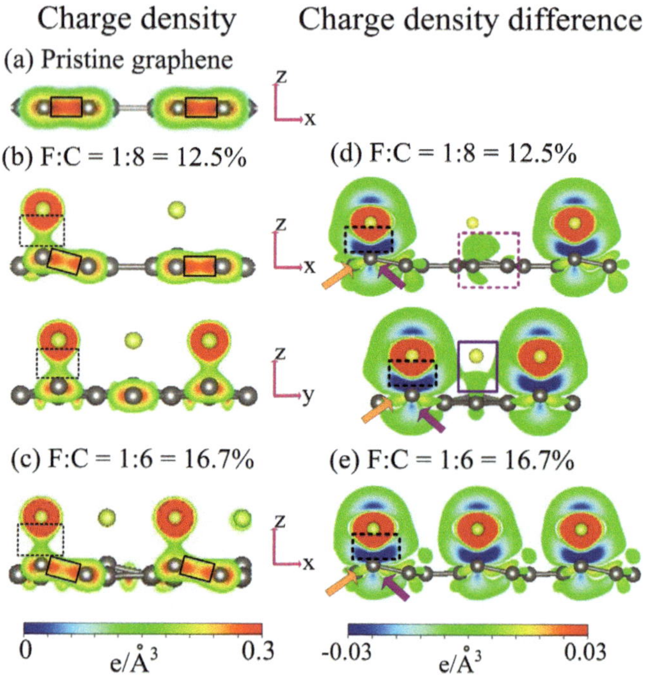

Figure 2.4 The spatial charge densities for (a) pristine graphene, (b) F:C = 1:8 = 12.5%, and (c) F:C = 1:6 = 16.7%. The corresponding charge density differences are, respectively, shown in (d) and (e).

the more complicated π and σ bands, and the seriously distorted Dirac cone (Figs. 2.2b–f). The obvious spatial distribution variations on (y, z) and (x, z) planes clearly illustrate the multi-orbital hybridizations of ($2p_x$, $2p_y$, $2p_z$) in F–C bonds. The sp^3 bonding resulted in a buckled graphene structure, which is indicated from the deformations of the σ bonding in the nearest C atom (orange arrows) and the π bonding in the next-nearest one (purple arrows). Between the non-passivated C atoms, the π and σ bondings almost remain the same as pristine graphene. Only slight deformations are presented as in the dashed purple rectangles (Fig. 2.4d).

The orbital-projected DOS can clearly illustrate the orbital hybridizations in F–F, C–F, and C–C bonds. For pristine graphene, the low-lying structures, as shown in the upper inset of Fig. 2.5a, are caused by the $2p_z$–$2p_z$ bondings among C atoms (red curve).

DOS presents a linear energy dependence near $E = 0$ and vanishes there, illustrating the characteristic of a zero-gap semiconductor. The π and π^* logarithmic-form peaks arise from the saddle points of parabolic bands near the M point (the dashed curve in Fig. 2.2a). Moreover, the shoulder structures, which are associated with the ($2p_x$, $2p_y$) orbitals, come to exist at deeper energy (~ -3 eV). They correspond to the extreme band-edge states of parabolic dispersions near the Γ point (Fig. 2.2a). The lower-energy DOS is dramatically altered after fluorination. Under the full fluorination (Fig. 2.5b), the V-shaped structure and two prominent symmetric peaks are absent, since the overall F–C interactions thoroughly destroy the π bondings of graphene. Instead, there exist an energy gap of 3.12 eV centered at E_F and several F-dominated special structures, a result of the F–F bondings. These clearly illustrate the coexistence of strong F–F, C–C, and F–C bonds. In addition, the $2p_z$ orbitals only make contributions at very deep energies (Fig. 2.5a). With the decrease in fluorination, there are more special structures in DOS, as indicated in Figs. 2.5b–f, being attributed to the narrower energy widths of F-dependent valence bands, the gradual recovery of carbon π bondings (red curves), and the partial contributions of F-$2p_z$ orbitals (blue curves).

The possession of remarkable properties enables use of fluori nated graphenes in many promising applications. They have been used in lithium-related batteries [40]. Edge-fluorinated graphene nanoplatelets could be used as high-performance electrodes for lithium-ion batteries and dye-sensitized solar cells [41]. Fluorinated graphenes could also be utilized as an electrode material for supercapacitors [42]. Moreover, the hybridized systems have potential performances in many other fields such as ink-jet printed technologies [43], the biological scaffold for promoting neuro-induction of stem cells [44], and ammonia detection [45].

2.3.2 Fluorinated 1D Graphene Nanoribbons

In this part, a detailed investigation of the geometric, electronic, and magnetic properties of fluorine-adatom-doped GNRs is carried out by various concentrations and adsorption positions of adatoms for both armchair and zigzag-edge-terminated shapes. The two-typical widths that can wholly represent for all AGNR and ZGNR

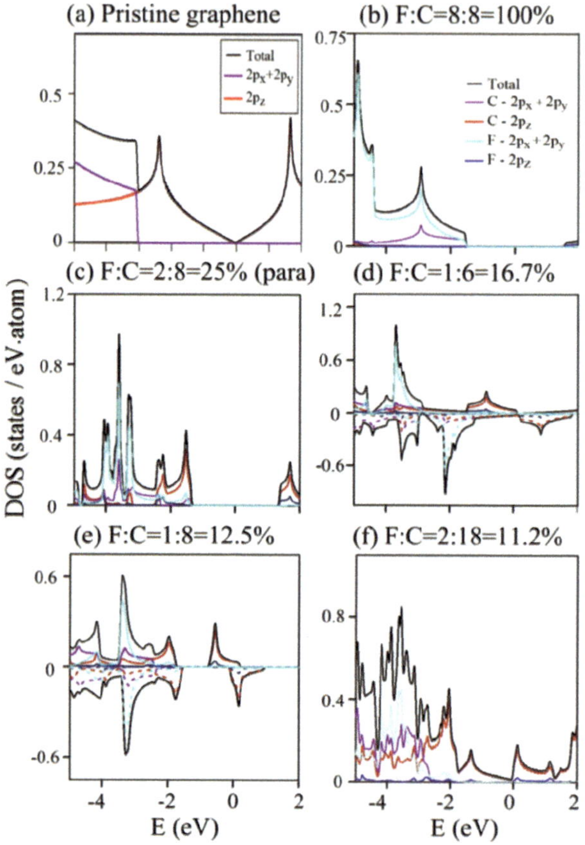

Figure 2.5 Orbital-projected DOS for (a) pristine graphene, (b) F:C = 8:8 = 100% (double-side), (c) F:C = 2:8 = 25% (para-site), (d) F:C = 1:6 = 16.7%, (e) F:C = 1:8 = 12.5%, (f) F:C = 2:18 = 11.2% (para-site).

are included in the studying model, as shown in Figs. 2.6a,b. The unit cells are described by the number of dimer lines and zigzag lines (N_A and N_Z) along y. The lattice constants along the periodic direction x, respectively, are $3b$ and $2\sqrt{3}b$ (b: C–C bond length) for AGNR and ZGNR. Besides, unnecessary dangling bonds in the GNRs are fully eliminated by hydrogen passivation. The geometric stability could be evaluated by the binding energy $E_b = (E_{tot}-E_{GNRs}-nE_A)/n$, in which E_{tot}, E_{GNRs}, and E_A account for the total energies of the combined system, pristine GNRs, and

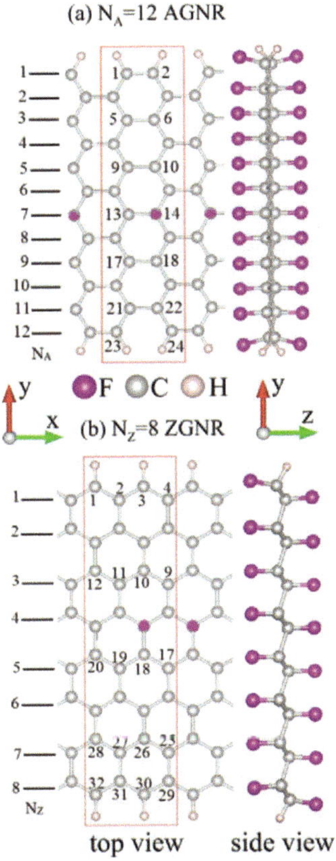

Figure 2.6 Geometric structures of F-adsorbed GNRs for (a) $N_{\mathrm{A}} = 12$ armchair and (b) $N_{\mathrm{Z}} = 8$ zigzag systems. The red rectangles represent the unit cells. The lattice constants are, respectively, $a = 3b$ and $a = 2\sqrt{3}b$ for armchair and zigzag GNRs. Numbers on the top of carbons denote the positions of adatoms.

the energy of all isolated adatoms; n displays for the number of adatoms, respectively. The fluorine-adatoms-adsorbed GNRs can be effectively synthesized from experiment since the evaluated binding energy is lower than −2.5 eV (Table 2.2). The top-site doping results in the optimized structures of halogen-adatoms-adsorbed GNRs regardless of any doping cases. The inhomogeneous buckled honeycomb lattice structures are revealed in the fluorine-adatoms-doped GNRs among other halogen-adatoms-adsorbed ones. C atoms

Table 2.2 Binding energy, Bohr magneton and magnetism, energy gap, number of holes in a unit cell, and geometric parameters for $N_A = 12$ armchair and $N_Z = 8$ zigzag GNRs under single- and double-side adsorption

GNRs	Adsorption configurations	E_b (eV)	Bohr magneton (μ_B)/magnetism	$E_g^{d(i)}$ (eV)/ Metal	Number of holes	F-C (Å)	C deviation (Å)	Nearest C-C (Å)
	Pristine	0	0/NM	$E_g^d = 0.6$	0	0	0	1.43
	(13)	−3.0145	0/NM	M	1	1.55	0.05	1.48
	(2)	−2.5155	0.47/FM	M	1	1.47	0.13	1.49
	$(5, 18)_s$	−2.6246	0/NM	M	2	1.55	0.055	1.49
	$(5, 18)_d$	−3.0145	0/NM	M	2	1.55	0.054	1.49
	$(2, 24)_s$	−3.3314	0.76/FM	M	1	1.47	0.132	1.49
	$(2, 24)_d$	−3.3526	0.76/FM	M	1	1.47	0.131	1.49
	$(2, 5)_d$	−4.2646	0/NM	$E_g^i = 0.64$	0	1.443	0.161	1.493
	$(C{:}F{=}\ 24:4)_d$	−3.1538	0/NM	M	2	1.503	0.097	1.491
	$(C{:}F{=}24{:}4)_d$	−3.3412	0.56/FM	M	1	1.454	0.15	1.497
AGNR $N_A = 12$	$(C{:}F{=}24{:}4)_d$	−4.2922	0/NM	$E_g^i = 0.65$	0	1.43	0.17	1.496
	$(C{:}F{=}24{:}6)_d$	−3.0107	0/NM	M	2	1.532	0.068	1.497
	$(C{:}F{=}24{:}6)_d$	−3.1625	0.57/FM	M	1	1.442	0.162	1.49
	$(C{:}F{=}24{:}6)_d$	−3.9762	0/NM	$E_g^i = 0.66$	0	1.413	0.185	1.51
	$(C{:}F{=}24{:}8)_d$	−3.4174	0/NM	M	2	1.446	0.154	1.5
	$(C{:}F{=}24{:}8)_d$	−3.3418	0.59/FM	M	1	1.472	0.127	1.5
	$(C{:}F{=}24{:}8)_d$	−3.8643	0/NM	$E_g^i = 0.72$	0	1.424	0.182	1.51

	(C:F=24:10)$_d$	−3.8426	0/NM	$E_g^i = 0.85$	0	1.425	0.182	1.51
	(C:F=24:14)$_d$	−3.8521	0/NM	$E_g^i = 2.25$	0	1.412	0.188	1.53
	(C:F=24:20)$_d$	−3.5124	0/NM	$E_g^i = 2.69$	0	1.417	0.193	1.53
	(C:F=24:24)$_d$	−3.6153	0/NM	$E_g^d = 3.2$	0	1.387	0.204	1.54
ZGNR $N_Z = 8$	Pristine	0	0/AFM	$E_g^d = 0.46$	0	0	0	1.43
	(3)	−4.5388	0.42/FM	M	1	1.46	0.139	1.49
	(14)	−2.5693	0.4/FM	M	1	1.55	0.06	1.48
	(3, 14)$_s$	−3.2856	0.4/FM	M	1	1.45	0.15	1.49
	(3, 14)$_d$	−3.4725	0.4/FM	M	1	1.45	0.15	1.49
	(10, 14)$_d$	−3.5521	0/AFM	$E_g^i = 0.2$	0	1.51	0.09	1.49
	(3, 30)$_s$	−4.2989	0/NM	$E_g^i = 0.46$	0	1.55	0.146	1.49
	(3, 30)$_d$	−4.6528	0/NM	$E_g^d = 0.46$	0	1.46	0.145	1.49
	(C:F=32:28)$_d$	−3.3645	0/NM	$E_g^d = 2.78$	0	1.38	0.219	1.52
	(C:F=32:32)$_d$	−3.4547	0/NM	$E_g^d = 3.1$	0	1.38	0.219	1.52

Note: NM, FM, and AFM correspond to non-magnetism, ferromagnetism, and anti-ferromagnetism, respectively.

closest to the F adatoms are deviated from the graphene plane, and they are sensitively dependent on concentrations and adsorption positions. For the single F adatom adsorption, the deviations of C atoms are 0.053 Å and 0.133 Å at the center and edge of GNR, respectively. They obviously increase with the adatom concentrations, e.g., 0.204 Å for highest adsorption concentration. The shortest and the longest F–C bond lengths (1.387 Å and 1.55 Å) correspond to the highest adsorption concentration and the ribbon-center-positioned single F adsorption. Likewise, the C–C bond lengths nearest to F adatoms are extended in the range of 0.05–0.11 Å as compared with those of pristine GNRs. This demonstrates the obvious change in the critical chemical bonds owing to the significant fluorination effects.

The honeycomb lattice symmetry, finite-size quantum confinement, and edge structure lead to many special 1D energy bands in the pristine AGNR (Fig. 2.7a). Specifically, there exists asymmetry between the occupied valence bands and the unoccupied conduction bands around the Fermi level ($E_{\rm F}$), where the direct band gap of $E_{\rm g}^{d} = 0.6$ eV emerges from the effect of the finite-size confinement at the Γ point. The π bonds of parallel $2p_z$ orbitals and the σ bonds of $(2p_x, 2p_y)$ orbitals are responsible for the electronic states of $E^{c,v} \leq 2$ eV and the deeper ones, as indicated in DOS (Fig. 2.11a). Most of the dispersions in the 1D band structures are parabolic, while a few of them have partially flat ones. All the energy dispersions depend on wave vectors monotonously except for the subband anti-crossings. The band-edge states that appear at $k_x = 0, 1$ (in unit of $\pi/3b$) and other states are related to subband anti-crossings that create the van Hove singularities in DOSs.

The energy bands are dramatically changed due to the adatom doping effects. For the single F-adatom-adsorbed AGNR, it presents metallic behavior with or without spin splitting, as shown in Figs. 2.7b,c. The Fermi level is shifted to valence energy bands, namely, a red shift. The free hole densities exist in the unoccupied valence states between two Fermi momenta ($\pm k_{\rm F}$ related to two valence bands intersecting with $E_{\rm F}$). The π bonds of C atoms create the low-lying energy bands, independent of adatom adsorption positions. Furthermore, the (F, C) co-dominated energy bands, accompanied with the σ bands, occur at $E^{v} \leq -2.5$ eV. They possess the weak

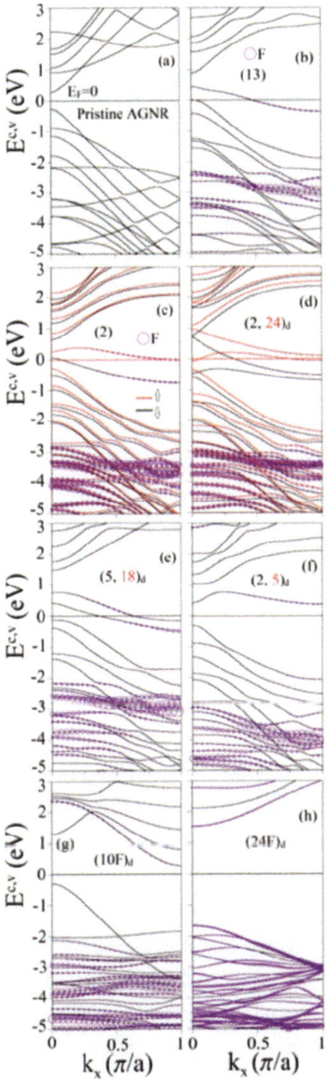

Figure 2.7 Band structures of $N_A = 12$ AGNR for (a) pristine, (b) (13)-, (c) (2)-, (d) $(2, 24)_d$-, (e) $(5, 18)_d$-, (f) $(2, 5)_d$-, (g) $(10F)_d$-, (h) (24F)-adsorption. Violet circles represent the contribution of F adatoms. The red and black curves denote the spin-split energy bands. [$(2, 24)_s$ and $(2, 24)_d$] stand for the single-side (black number; subscript s) and double-side (black and red number; subscript d) adsorption positions, $(10F)_d$ and $(24F)_d$ represent the double-side adsorption with 10 and 24 adatoms, respectively.

energy dispersions or the narrow energy band widths. The critical F–C bonds and the significant modifications of π and σ bands can be characterized by the special features of 1D energy bands.

The variations in the concentration, adsorption position, and edge structure can create the obvious change in the 1D energy bands. With the gradual rise in adatom concentration, the Fermi-momentum states are drastically varied, as shown in Figs. 2.7d–f for two-F adsorptions. The total free hole densities become higher [$(5,\ 18)_d$ in Fig. 2.7e; $(5,\ 18)_s$ in Table 2.2] or remain the same [$(2,\ 24)_d$ in Fig. 2.7d; $(2,\ 24)_s$ in Table 2.2], compared to that of the single adatom adsorption (Figs. 2.7b,c). Note that the expressions [$(5,\ 18)_s$ and $(5,\ 18)_d$ or $(2,\ 24)_s$ and $(2,\ 24)_d$] stand for the adatom adsorption positions under the single-side (black number; subscript s) and double-side (black and red number; subscript d) adsorption. Furthermore, the metallic 1D band structures might be thoroughly turned into the semiconducting one for the very close adatom–adatom distances, e.g., an indirect gap of $E_{\mathrm{g}}^{i} = 0.64$ eV for the $(2,\ 5)_d$ adsorption (Fig. 2.7f). This exhibits the termination of the extended π bonds in AGNRs. Specifically, most of the F-doped AGNRs become the large-gap semiconductors under the efficiently high F adsorption concentrations (Figs. 2.7g,h). There exist more F-dominated energy bands that are located in the valence bands. The energy bands determine the magnitude of energy gap, and E_{g} also depends on the σ bonding of carbon atoms (DOSs in Figs. 2.11b,c). Under the 100% adsorption, the non-magnetic wide-gap semiconductor of $E_{\mathrm{g}}^{i} = 3.2$ eV is generated (Fig. 2.7h).

A clear examination of the free hole densities is worthy of the detailed calculations. Such free hole densities are expressed by the relation $\lambda = 2k_{\mathrm{F}}/\pi$, and linearly proportional to the Fermi momentum for each partially unoccupied valence band. There are no linear relations between λ and F adatom concentration except for the single F adatom adsorption. The single F-adatom-doped AGNR can only create one free hole for any adsorption positions (Table 2.2), i.e., one electron is absorbed by a fluorine adatom from the bonded carbon atom due to the strong electron affinity. The metallic 4-, 6-, and 8-adatom adsorptions can create two or one free holes in a unit cell corresponding to the spin-degenerate and spin-split energy bands, respectively (Table 2.2). The F-adatoms-adsorbed GNRs are

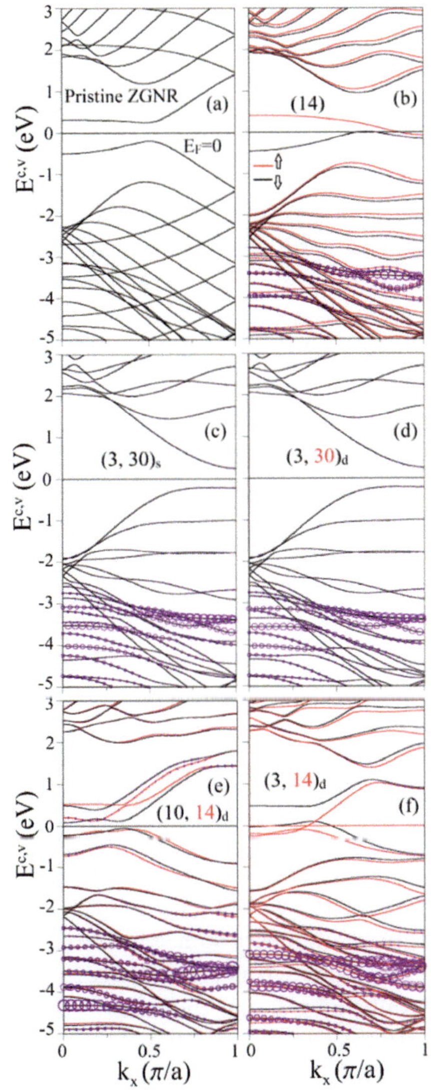

Figure 2.8 Band structures of $N_Z = 8$ ZGNR for (a) pristine, (b) (14)-, (c) $(3, 30)_s$-, (d) $(3, 30)_d$-, (e) $(10, 14)_d$-, and (f) $(3, 14)_d$-adatom adsorption. Violet circles represent the contribution of F adatoms. The red and black curves denote the spin-split energy bands. The subscripts s and d, respectively, correspond to the single- and double-side adsorptions, accompanied with the black number and (black, red) numbers.

in sharp contrast with the alkali-adatoms-doped ones. The latter belongs to n-type metals even for the 100% adatom concentration. Independent of adatom distributions, each alkali adatom generates one conduction electron from the outermost s orbital by means of the significant alkali-C bond. This makes conduction electrons very high.

The electronic and magnetic properties are diversified by the various edge structures. There exists a major difference in the energy bands of AGNRs and ZGNRs. Pristine ZGNRs possess a pair of partially flat valence and conduction bands nearest to E_F at small k_x's (Fig. 2.8a), corresponding to wave functions localized at the zigzag boundaries [46]. Moreover, these energy bands also exhibit the double degeneracy for the spin degree of freedom even if they are closely related to the anti-ferromagnetic configuration across the ribbon center and the ferromagnetic one at the same edge (Fig. 2.9d). Besides, the band-edge states of $N_Z = 8$ ZGNR create a direct gap of $E_g^d = 0.46$ eV. The partially flat 1D band structures with the localized charge distributions could be changed by F adatom doping such as energy dispersions, energy gap, and state degeneracy (Figs. 2.8b–f). For the single F-adatom-adsorbed ZGNRs, the number of edge-localized energy bands is reduced to a half of the spin splitting (Fig. 2.8b). Furthermore, the energy bands intersect with the Fermi level that exhibits metallic behavior. Such behavior vanishes under the adsorption of two-F adatoms, which locate at both zigzag edges [$(3, 30)_s$ in Fig. 2.8c and $(3, 30)_d$ in Fig. 2.8d], leading to a direct-gap semiconductor (Table 2.2). Under the central adsorption of two-F adatoms, the behavior of an indirect narrow-gap semiconductor is presented with spin splitting (Fig. 2.8e). Especially, the partially flat semiconducting band structures fully turn into the spin-splitting metallic ones under the adsorption of two-F adatoms, which situate at the edge and center [$(3, 14)_d$ in Fig. 2.8f].

There exist five kinds of spin-dependent electronic and magnetic configurations. The pristine AGNRs are confinement-induced semiconductors without spin-split energy bands and magnetism (the first kind in Fig. 2.7a and Table 2.2). The similar semiconducting properties are also revealed in the F-adatoms-doped AGNRs (Figs. 2.7f–h). Nevertheless, the metallic systems might present the spin-degenerate energy bands without magnetism (the second

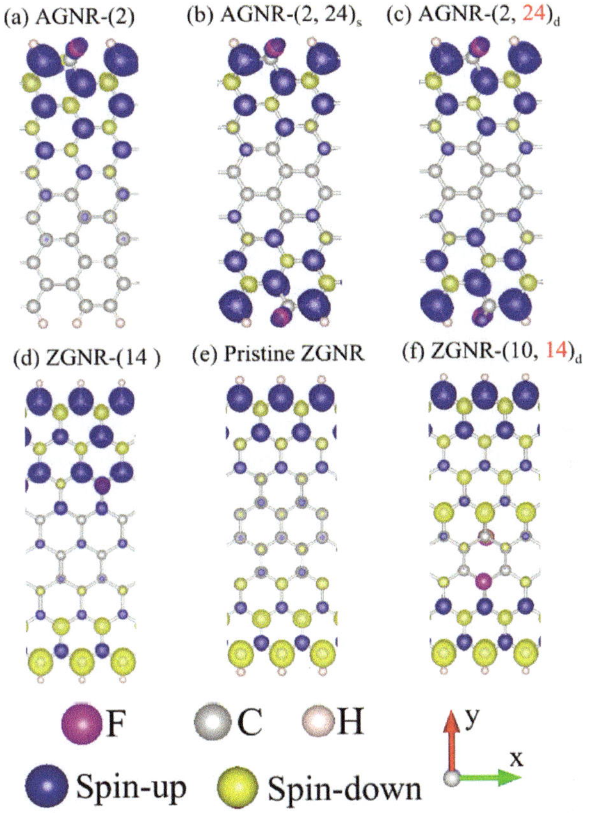

Figure 2.9 Spin density of $N_A = 12$ AGNR for (a) (2)-, (b) $(2, 24)_s$-, (c) $(2, 24)_d$-adatom adsorption, and $N_Z = 8$ ZGNR for (d) (14)-, (e) pristine, and (f) $(10, 14)_d$-adatom adsorption.

kind in Figs. 2.7b,e), or the spin splitting with the ferromagnetic configuration (the third kind in Figs. 2.7c,d). On the other hand, the pristine ZGNRs are the anti-ferromagnetic semiconductors without spin splitting (the fourth kind in Fig. 2.8a). Furthermore, the F-adatoms-doped ones present the first kind (Figs. 2.8c,d), the third kind (Figs. 2.8b,f), or the fifth kind (the semiconducting behavior with the anti-ferromagnetic configuration (Fig. 2.8e)).

The electronic and magnetic configurations could be clearly interpreted by the spatial spin densities and magnetic moments. Third, fourth, and fifth kinds of electronic and magnetic configurations exhibit the different arrangements (Figs. 2.9a–d, Fig. 2.9e, and

Fig. 2.9f, respectively), in which the competition between the spin-up and spin-down configurations creates the net magnetic moment in a unit cell (Table 2.2). The spin densities are mostly distributed near the edge structures except for the fifth kind (Fig. 2.9f). When only one F adatom is adsorbed at the edge of AGNR, this leads to the ferromagnetic spin-up configuration, as shown in Fig. 2.9a, and results in a net moment of 0.47 μ_B. For two-F adsorptions near both boundaries, AGNRs exhibit the enhanced ferromagnetism across the ribbon center with 0.76 μ_B (Figs. 2.9b,c), while magnetism is absent in ZGNRs (Figs. 2.9e,f). These illustrate that F adatoms close to the armchair and zigzag edges, respectively, create and destroy the same-spin arrangement on the spot. Specifically, two F adatoms adsorption at the edge and center of ZGNR creates an unusual anti-ferromagnetic configuration with a zero magnetic moment (Fig. 2.9f).

The multi-orbital hybridizations are analyzed by the spatial charge density (ρ), the charge density difference ($\Delta\rho$), and the partial charge density (ρ_P). ($\Delta\rho$) is calculated by subtracting the charge density of GNRs and F adatoms from that of F-adatom-doped system. Also the chemical bonds as well as the charge transfer are clearly illustrated by (ρ). The π and σ bonds in the planar GNRs are formed by the parallel 2p$_z$ orbitals and the planar (2p$_x$, 2p$_y$) ones, as shown in Fig. 2.10a for $N_A = 12$ AGNR (the solid and dashed rectangles). The high charge density between F and C could have obvious changes in π bonding and observable reductions in σ bonding (Figs. 2.10b,d). The strong fluorination effects are also revealed in drastic density variations, especially for ($\Delta\rho$), near F adatoms on (x, z) and (y, z) planes (Figs. 2.10c,e). These clearly indicate the complicated (2p$_x$, 2p$_y$, 2p$_z$)-(2p$_x$, 2p$_y$, 2p$_z$) hybridizations in F–C bonds. When the distance between two fluorine adatoms is sufficiently short, there exist significant (2p$_x$, 2p$_y$) hybridizations in F–F bonds, as illustrated in Figs. 2.10c,e (red rectangles on (x, z) and (y, z) planes). As for the metallic and semiconducting behaviors, they are characterized by the distorted π bondings in the partial charge density related to electronic states very close to E_F (Figs. 2.10f,h and Figs. 2.10g,i).

The diverse electronic and magnetic configurations can be verified by the orbital-projected DOSs (Fig. 2.11). There exhibit

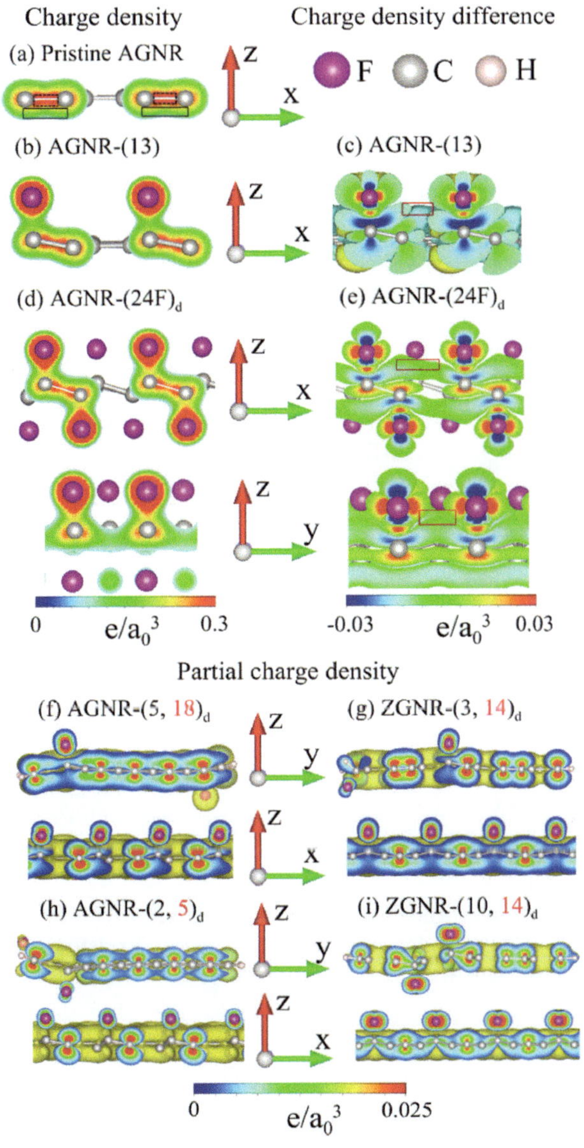

Figure 2.10 Spatial charge density of $N_A = 12$ AGNR for (a) pristine, (b) (13)-, and (d) $(24F)_d$-adatom adsorption; charge density difference of $N_A = 12$ AGNR for (c) (13)-, (e) $(24F)_d$-adatom adsorption. Partial charge density is shown for (f) AGNR-$(5, 18)_d$-, (g) ZGNR-$(3, 14)_d$-, (h) AGNR-$(2, 5)_d$-, and (i) ZGNR-$(10, 14)_d$-adatom adsorption.

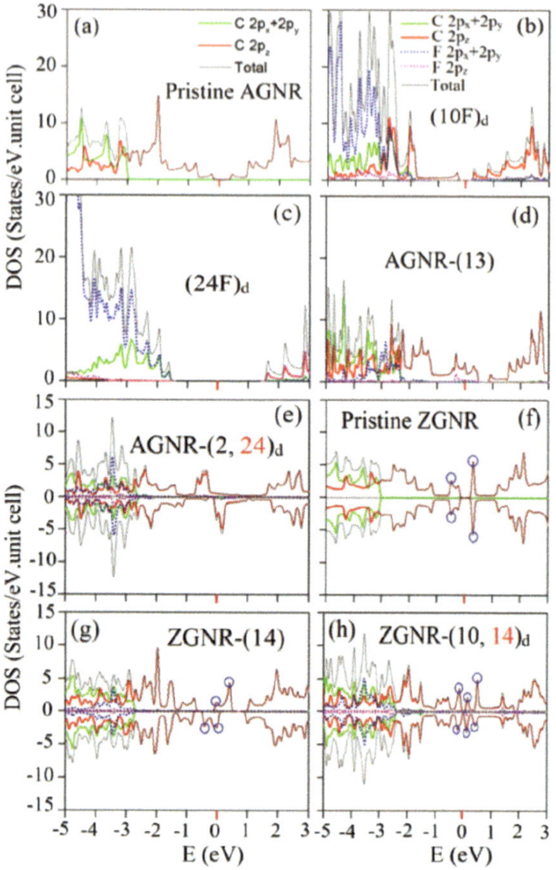

Figure 2.11 Orbital-projected DOSs for (a) pristine, (b) $(10F)_d$-, (c) $(24F)_d$-, (d) (13)-, (e) $(2, 24)_d$-adsorbed $N_A = 12$ AGNR; (f) pristine; (g) (14)-, (h) $(10, 14)_d$-adsorbed $N_Z = 8$ ZGNR. Blue circles correspond to the partially flat bands.

many special structures, in which the asymmetric and symmetric peaks emerge from the parabolic and partially flat energy bands. When the adatom concentration is lower than 50% (Figs. 2.11b,d,e), the low-energy DOS is dominated by the π bonding of C-2p$_z$ orbitals (red curves). Such bonding also makes contributions to the deeper-energy DOS. The peak structures owing to the σ bonding of (2p$_x$, 2p$_y$) orbitals occur at $E < -2.5$ eV (green curves). Similar features take place with respect to the ($2p_x$, $2p_y$), and $2p_z$ orbitals of the F adatoms (dashed blue and pink curves). The former possess a sufficiently wide energy width of 2 eV, so that there exist significant ($2p_x$, $2p_y$) orbital hybridizations in F–F bonds. All the orbitals can create the merged peak structures at deeper energy, clearly illustrating the (($2p_x$, $2p_y$, $2p_z$)-($2p_x$, $2p_y$, $2p_z$)) multi-orbital hybridizations in F–C bonds. Specifically, the energy width of the bands might be more than 5 eV under the highest adatom concentration (Fig. 2.11c). Furthermore, the corresponding peaks are stronger than those of the σ bands. The occupied valence bands are closely related to the ($2p_x$, $2p_y$) orbitals of F and C, especially for the orbital hybridization in F–F bonds.

Five kinds of electronic and magnetic configurations are characterized by the specific peak structures near the Fermi level. A pair of anti-symmetric peaks near $E = 0$, which are divergent in the opposite direction, represents an energy gap in the absence of spin splitting (the first kind), as shown in Figs. 2.11a–c. But for non-magnetic metals (the second kind), the similar pair presents a blue shift $(0.5 - 1.0)$ eV (Fig. 2.11d), and DOS is finite at the Fermi level. The spin-polarized peak structures are revealed in ferromagnetic metals (the third kind in Figs. 2.11c,g). The low-energy peaks are quite different for spin-up and spin-down configurations, in which they are asymmetric about $E = 0$, and the former predominates the occupied states of $E < 0$. Specifically, the partially flat energy bands in a pristine ZGNR can create a pair of symmetric peaks (blue circles in Fig. 2.11f), accompanied with an energy gap and one peak due to an extra band-edge state (Fig. 2.8a). This corresponds to an anti-ferromagnetic semiconductor with spin degeneracy (the fourth kind). There are more pairs of symmetric peaks centered about $E = 0$ in the presence of spin splitting, as indicated in Fig. 2.11h. The fifth kind of peak structure arises from the narrow-gap semiconducting band structure of two F-adatoms-absorbed ZGNRs with anti-ferromagnetic behavior.

2.3.3 Experimental Measurements

On the experimental aspects, a scanning tunneling microscope (STM), an useful tool for imaging surfaces at atomic level with a good resolution, has been successfully utilized to identify the unique geometric structures of graphene-related systems, including graphite, graphene, graphene compounds, carbon nanotubes, and GNRs. The atomic-scaled observations clearly reveal the 2D networks of local defects [47], the buckled and rippled structures of graphene islands [48], the adatom distributions on graphene surface [49], the nanoscale width of GNR [50], and the chiral arrangements of the hexagons on the planar edges [51] and a cylindrical surface [52]. For fluorinated 2D and 1D graphene, the buckled structure, the adatom height, and the bond length (Tables 2.1 and 2.2) are worthy of further STM verifications. Such measurements are very useful in identifying the significant multi-orbital hybridizations in F–C, F–F, and C–C chemical bonds.

Angle-resolved photoemission spectroscopy (ARPES), a direct experimental technique to observe the distribution of electrons in the reciprocal space of solids, is one of the most powerful experimental methods to examine the wave-vector-dependent electronic structures. Besides, the feature-rich electronic structures within the distinct dimensions are revealed in the experimental measurements on graphene-related systems. For example, the verified energy bands include an isotropic Dirac-cone structure with linear energy dispersions in monolayer graphene [53], two pairs of parabolic bands in bilayer AB stacking [54], the bilayer- and monolayer-like energy dispersions, respectively, at $k_z = 0$ and zone boundary in AB-stacked graphite [55] and 1D parabolic energy bands with a direct energy gap in AGNRs [56]. In addition, an edge-localized partially flat band is deduced to have an association with the zigzag-like steps on graphite surface. The measurement on fluorinated graphene shows that the Fermi level is shifted 0.79 eV below the Dirac point, indicating the p-type doping. Up to now, the ARPES measurements on the adatom-enriched 1D energy bands of GNRs are absent. The ARPES and spin-resolved ARPES [57] are available in verifying five kinds of predicted electronic and magnetic configurations in F-adatom-absorbed GNRs.

Scanning tunneling spectroscopy (STS), an extension of STM, is used to provide information about the density of electrons in a sample as a function of their energy, with the tunneling differential conductance (dI/dV) proportional to DOS, and can serve as a very efficient method to identify the dimension-enriched special structures in DOS. The measured DOSs show the splitting π and π^* peaks and a finite value near the E_F characteristic of the semi-metallic behavior in graphite [58]; a linear E-dependence vanishing at the Dirac point in monolayer graphene; the asymmetry-created peak structures in bilayer graphene [59]; a prominent peak at E_F arising from partially flat bands in tri-layer and penta-layer ABC-stacked graphene [60]; the geometry-dependent energy gaps; and the asymmetric peaks of 1D parabolic bands in carbon nanotubes [61] and GNRs. The STS and spin-resolved STS can be utilized to examine the orbital-projected DOSs of F-adatom-absorbed 2D graphene and 1D GNRs, covering the finite value at E_F, energy gap, and spin-polarized peak structures.

2.4 Concluding Remarks

The geometric structures and the electronic and magnetic properties of F-adatom-adsorbed 2D graphene and 1D GNRs are investigated using the first-principles calculations. Discussions on the adatom-dominated band structure, the spin arrangement, spatial charge density, and the orbital-projected DOS are useful to elucidate the feature-rich electronic and magnetic properties. Similar analyses can be further generalized to the emergent 2D and 1D layered materials, with nanoscale thickness and unique lattice symmetries, covering 2D and 1D silicene, germanene, tinene, phosphorene, MoS_2, and so on. The fluorination-induced diverse effects mainly arise from the complicated relations among 2D and 1D honeycomb lattice symmetries, the top-site F–C multi-orbital hybridizations, finite-size quantum confinements, and specific edge structures. The metallic (with/without spin-splitting) or large-gap semiconducting behaviors can turn into the highly potential applications, such as electronic, optical, and spintronic devices.

Each F-adatom-adsorbed 2D and 1D graphene can create an inhomogeneous buckled structure, in which the adatom height and the change in the C–C bond length are mainly determined by the significant fluorination effect. The multi-orbital hybridizations in F–C, C–C, and F–F bonds of fluorinated graphene lead either to the metals (with/without magnetism) or to the large-gap semiconductors (without magnetism), depending on the adatom concentrations and distributions. The carbon-dominated magnetisms are revealed in the fluorination of 12.5% and 16.7%. Furthermore, the critical multi-orbital hybridizations and the edge-dependent spin distributions of the fluorinated GNRs are responsible for five kinds of electronic and magnetic properties, covering non-magnetic and ferromagnetic metals, non-magnetic semiconductors, and anti-ferromagnetic semiconductors with/without spin splitting. Besides, F-adatom-doped AGNRs can create one or two holes per unit cell corresponding to the spin-split and spin-degenerate π-electronic energy bands (non-magnetism and ferromagnetism), respectively. However, F-adatom-doped ZGNRs only exhibit the case with lower carrier density. The predicted geometric, electronic, and magnetic properties of fluorinated 2D and 1D graphene can be verified by STM, ARPES, and STS, respectively.

Acknowledgments

This research is funded by the Vietnam National Foundation for Science and Technology Development (NAFOSTED) under grant number 103.01-2017.74. Also we would like to thank the Department of Physics, National Cheng Kung University, Taiwan for the computational support.

References

1. S. Jandhyala, G. Mordi, B. Lee, G. Lee, C. Floresca, P. R. Cha, et al., *ACS Nano*, 2012, **6**, pp. 2722–2730.
2. D. Wei, Y. Liu, Y. Wang, H. Zhang, L. Huang, and G. Yu, *Nano Letters*, 2009, **9**, pp. 1752–1758.
3. N. T. T. Tran, S. Y. Lin, C. Y. Lin, and M. F. Lin, *CRC Press, Catalog no. K43309*, 2017, "Geometric and Electronic Properties of Graphene-Related Systems: Chemical Bonding Schemes," p. 194, ISBN: 978-1-1385-5652-2.

4. K. Guerin, J. P. Pinheiro, M. Dubois, Z. Fawal, F. Masin, R. Yazami, et al., *Chemistry of Materials*, 2004, **16**, pp. 1786–1792.
5. M. Cai, D. Thorpe, D. H. Adamson, and H. C. Schniepp, *Journal of Materials Chemistry*, 2012, **22**, pp. 24992–25002.
6. W. Feng, P. Long, Y. Feng, and Y. Li, *Advanced Science*, 2016, **3**, 1500413.
7. L. Jiao, X. Wang, G. Diankov, H. Wang, and H. Dai, *Nature Nanotechnology*, 2010, **5**, pp. 321–325.
8. L. Talirz, H. Söde, J. Cai, P. Ruffieux, S. Blankenburg, R. Jafaar, et al., *Journal of the American Chemical Society*, 2013, **135**, pp. 2060-2063.
9. M. Y. Han, B. Özyilmaz, Y. Zhang, and P. Kim, *Physical Review Letters*, 2007, **98**, 206805.
10. Z. Chen, Y. M. Lin, M. J. Rooks, and P. Avouris, *Physica E: Low-Dimensional Systems and Nanostructures*, 2007, **40**, pp. 228–232.
11. X. Li, X. Wang, L. Zhang, S. Lee, and H. Dai, *Science*, 2008, **319**, pp. 1229–1232.
12. S. S. Datta, D. R. Strachan, S. M. Khamis, and A. C. Johnson, *Nano Letters*, 2008, **8**, pp. 1912–1915.
13. D. V. Kosynkin, A. L. Higginbotham, A. Sinitskii, J. R. Lomeda, A. Dimiev, B. K. Price, et al., *Nature*, 2009, **458**, pp. 872–876.
14. V. Barone, O. Hod, and G. E. Scuseria, *Nano Letters*, 2006, **6**, pp. 2748–2754.
15. Q. Yan, B. Huang, J. Yu, F. Zheng, J. Zang, J. Wu, et al., *Nano Letters*, 2007, **7**, pp. 1469–1473.
16. Y. Pak, S. M. Kim, H. Jeong, C. G. Kang, J. S. Park, H. Song, et al., *ACS Applied Materials and Interfaces*, 2014, **6**, pp. 13293–13298.
17. C. P. Chang, B. R. Wu, R. B. Chen, and M. F. Lin, *Journal of Applied Physics*, 2007, **101**, 063506.
18. S. Y. Lin, S. L. Chang, F. L Shyu, J. M. Lu, and M. F. Lin, *Carbon*, 2015, **86**, pp. 207–216.
19. Y. C. Huang, C. P. Chang, and M. F. Lin, *Physical Review B*, 2008, **78**, 115422.
20. S. L. Chang, B. R. Wu, J. H. Wong, and M. F. Lin, *Carbon*, 2014, **77**, pp. 1031–1039.
21. C. Y. Lin, J. Y. Wu, Y. J. Ou, Y. H. Chiu, and M. F. Lin, *Physical Chemistry Chemical Physics*, 2015, **17**, pp. 26008–26035.
22. S. L. Chang, S. Y. Lin, S. K. Lin, C. H. Lee, and M. F. Lin, *Scientific Reports*, 2014, **4**, 6038.

23. Y. T. Lin, H. C. Chung, P. H. Yang, S. Y. Lin, and M. F. Lin, *Physical Chemistry Chemical Physics*, 2015, **17**, pp. 16545–16552.
24. H. Sadeghi, M. T. Ahmadi, S. M. Mousavi, R. Ismail, and M. H. Ghadiry, *Modern Physics Letters B*, 2012, **26**, 1250047.
25. C. P. Chang, Y. C. Huang, C. L. Lu, J. H. Ho, T. S. Li, and M. F. Lin, *Carbon*, 2006, **44**, pp. 508–515.
26. E. J. Kan, Z. Li, J. Yang, and J. G Hou, *Applied Physics Letters*, 2007, **91**, 243116.
27. M. F. Lin and F. L. Shyu, *Journal of the Physical Society of Japan*, 2000, **69**, pp. 3529–3532.
28. Y. C. Huang, C. P. Chang, and M. F. Lin, *Nanotechnology*, 2007, **18**, 495401.
29. Y. Li, Z. Zhou, P. Shen, and Z. Chen, *ACS Nano*, 2009, **3**, pp. 1952–1958.
30. D. K. Nguyen, Y. T. Lin, S. Y. Lin, Y. H. Chiu, N. T. T. Tran, and M. F. Lin, *Physical Chemistry Chemical Physics*, 2017, **19**, pp. 20667–20676.
31. D. Krepel and O. Hod, *Surface Science*, 2011, **605**, pp. 1633–1642.
32. Y. T. Lin, S. Y. Lin, Y. H. Chiu, and M. F. Lin, *Scientific Reports*, 2017, **7**, 1722.
33. C. G. da Rocha, P. A. Clayborne, P. Koskinen, and H. Häkkinen, *Physical Chemistry Chemical Physics*, 2014, **16**, pp. 3558–3565.
34. G. Kresse and J. Furthmüller, 2001, Vienna ab-initio simulation package (VASP), Vienna: Vienna University.
35. J. P. Perdew and K. Burke, *Physical Review Letters*, 1996, **77**, 3865.
36. P. E. Blöchl, *Physical Review B* , 1994, **50**, 17953.
37. N. T. T. Tran, D. K. Nguyen, O. E. Glukhova, and M. F. Lin, *Scientific Reports*, 2017, **7**, 17858.
38. O. Leenaerts, H. Peelaers, A. D. Hermández-Nieves, B. Partoens, and F. M. Peeters, *Physical Review B*, 2010, **82**, 195436.
39. J. T. Robinson, J. S. Burgess, C. E. Junkermeier, S. C. Badescu, T. L. Reinecke, F. K. Perkins, et al., *Nano Letters*, 2010, **10**, pp. 3001–3005.
40. I. Y. Jeon, M. J. Ju, J. Xu, H. J. Choi, J. M. Seo, M. J. Kim, et al., *Advanced Functional Materials*, 2015, **25**, pp. 1328–1328.
41. C. Sun, Y. Feng, C. Qin, Q. Zhang, and W. Feng, *Nanoscale*, 2014, **6**, pp. 2634–2641.
42. D. A. Siddhanti, D. J. Nash, M. A. Navarro, D. M. Mills, A. Khaniya, B. Dhar, et al., *Journal of Materials Science*, 2017, **52**(20), pp. 11977–11987.
43. M. H. Ervin, L. T. Le, and W. Y. Lee, *Electrochimica Acta*, 2014, **147**, pp. 610–616.
44. D. D. Chronopoulos, A. Bakandritsos, M. Pykal, R. Zboril, and M. Otyepka, *Applied Materials Today*, 2017, **9**, pp. 60–70.

45. L. Liu, M. Zhang, Z. Xiong, D. Hu, G. Wu, and P. Chen, *Carbon*, 2014, **81**, pp. 702–709.

46. G. Lee and K. Cho, *Physical Review B*, 2009, **79**, 165440.

47. J. Cervenka, M. I. Katsnelson, and C. F. J. Flipse, *Nature Physics*, 2009, **5**, pp. 840–844.

48. L. Meng, W. Y. He, H. Zheng, M. Liu, H. Yan, W. Yan, et al., *Physical Review B*, 2013, **87**, 205405.

49. D. Pandey, R. Reifenberger, and R. Piner, *Surface Science*, 2008, **602**, pp. 1607–1613.

50. P. Ruffieux, J. Cai, N. C. Plumb, L. Patthey, D. Prezzi, A. Ferretti, et al., *ACS Nano*, 2012, **6**, pp. 6930–6935.

51. C. Tao, L. Jiao, O. V. Yazyev, Y. C. Chen, J. Feng, and X. Zhang, *Nature Physics*, 2011, **7**, pp. 616–620.

52. J. W. Wilder, L. C. Venema, A. G. Rinzler, R. E. Smalley, and C. Dekker, *Nature*, 1998, **391**, pp. 59–62.

53. T. Ohta, A. Bostwick, J. L. McChesney, T. Seyller, K. Horn, and E. Rotenberg, *Physical Review Letters*, 2007, **98**, 206802.

54. T. Ohta, A. Bostwick, T. Seyller, K. Horn, and E. Rotenberg, *Science*, 2006, **313**, pp. 951–954.

55. A. Grüneis, C. Attaccalite, T. Pichler, V. Zabolotnyy, H. Shiozawa, S. L. Molodtsov, et al., *Physical Review Letters*, 2008, **100**, 037601.

56. K. Sugawara, T. Sato, S. Souma, T. Takahashi, and H. Suematsu, *Physical Review B*, 2006, **73**, 045124.

57. D. Usachov, A. Fedorov, M. M. Otrokov, A. Chikina, O. Vilkov, A. Petukhov, et al., *Nano Letters*, 2015, **15**, pp. 2396–2401.

58. Z. Klusek, *Applied Surface Science*, 1999, **151**, pp. 251–261.

59. A. Luican, G. Li, A. Reina, J. Kong, R. R. Nair, K. S. Novoselov, et al., *Physical Review Letters*, 2011, **106**, 126802.

60. Y. Que, W. Xiao, H. Chen, D. Wang, S. Du, and H. J. Gao, *Applied Physics Letters*, 2015, **107**, 263101.

61. T. W. Odom, J. L. Huang, P. Kim, and C. M. Lieber, *Nature*, 1998, **391**, pp. 62–64.

Chapter 3

Perspectives of New 2D Films Based on Covalently Bonded Graphene and Carbon Nanotubes for Electronic Applications

Gang Zhang[a] **and Olga E. Glukhova**[b,c]

[a]*Agency for Science, Technology and Research, Institute of High Performance Computing, 1 Fusionopolis Way, 138632, Singapore*
[b]*Physical Department, Saratov State University, 83 Astrakhanskaya Street 410010, Saratov, Russia*
[c]*I. M. Sechenov First Moscow State Medical University, Laboratory of Biomedical Nanotechnology, Trubetskaya Street 8-2, Moscow, 119991, Russia*
glukhovaoe@info.sgu.ru

This chapter is devoted to the study of the electronic structure and electrical properties of two-dimensional (2D) films based on carbon nanotube/graphene composite with horizontal packing of tubes between graphene sheets. The influence of topology—the chirality of nanotubes and the distance between them—on the electron energy and electrical characteristics of the composites is studied. To reveal new physical regularities of the hybrid carbon nanostructures under consideration, numerical experiments have been performed using quantum mechanical and molecular dynamics methods.

2D and 3D Graphene Nanocomposites: Fundamentals, Design, and Devices
Edited by Olga E. Glukhova

ISBN 978-981-4800-41-9 (Hardcover), 978-0-429-20150-9 (eBook)
www.jennystanford.com

3.1 Introduction

Hybrid nanocomposites based on 2D graphene and single-walled carbon nanotubes (SWCNTs) are promising new materials whose properties and possibilities are not fully discovered. To date, synthesized CNT/graphene composites can be divided into two large morphological groups [1]: (1) three-dimensional (3D) composite "pillared graphene" in which tubes are vertically oriented on graphene layers and located between them as if "growing" from one graphene layer and spiraling into another (both seamless and ohmic contacts between graphene and CNT are possible); (2) 2D hybrid films in which tubes are located between adjacent graphene layers (CNTs are bonded to graphene covalently or by van der Waals forces). The second group consists of structures with a parallel, chess, and chaotic arrangement of CNTs to each other. Methods for the synthesis of pillared graphene are different. The most known method is chemical vapor deposition (CVD) where graphene is deposited with catalytic particles on which CNTs are grown [2, 3]. Hybrid 2D films are also synthesized by the CVD method with additional techniques for transferring polymer from CNT to graphene [4].

The functionality of the graphene/CNT composite has been already proved by numerous experimental studies and prototypes of new nanodevices designed on its base. Pillared graphene with a seamless CNT–graphene contact is already regarded as the new supercapacitor [5–7]. If it is doped with nitrogen and sulfur atoms, it can be an effective filler for metal–air batteries, increasing the potential by almost half compared to an unalloyed composite [8]. The 3D composite functionalized by oxygen has also been proposed as an independent nanoelectronic device—an ultrafast compact energy store where positively infected nanotubes act as anode and a negatively charged oxidized graphene acts as cathode [9]. When functionalized with tin oxide or molybdenum disulfide, a 3D composite significantly improves the efficiency of lithium-ion batteries [2, 10].

The hybrid 2D film exhibits high performance as the photo-sensitive element of a photodetector in the 400–1500 nm range: one absorbed by film photon induces transport of 10^5 electrons with a response time of ~100 μs [11]. Also this film is effective as a miniature energy source [1, 12]. Functionalization of tubes by OH groups makes the 2D composite hydrophilic, which opens new possibilities for its application in inkjet printing and high-conductivity polymers [13]. The hybrid 2D film modified by manganese oxide with multi-walled CNTs increases the selective sensory properties of a material to hydrogen peroxide [14].

Modern synthetic technologies make it possible not only to plan synthesis but also to regulate the formation of a supercomposite layer by layer. An example is a hybrid carbon 2D film in which CNTs between layers of graphene are "stacked" in regular chessboard order [15]. Possible physical and chemical phenomena in such topological grids and possibilities of application are now being actively investigated. One of the most expected applications is an efficient material for the design of solar cells [16]. Also such supercomposites can be used as a flexible and miniature electric "carpet" for local heating in micro- and nanodevices [17]. It is explained by the unique heat-conducting properties of the supercomposite, which are just beginning to be studied. Functionalization of supercomposites with irregular stacking of CNTs (in the chessboard order or even chaotic) between graphene sheets by GaN makes it possible to obtain a new class of promising light-emitting diodes on the basis of this material [7].

The scientific significance of new materials is obvious; the discovery of unique properties and applications pushes the horizons of human capabilities and allows solving many life support problems. The above-mentioned graphene composites clearly demonstrate the relevance and importance of searching for new materials and studying their properties. From the wide variety of potential applications of hybrid CNT graphene composites of a new generation, it can be concluded that many important aspects have not been investigated: electronic and optical properties, topological regularities of CNTs between graphene layers, and the influence of topology on properties.

This chapter is devoted to the investigation of the in silico regularities in the topology of the 2D composite material SWCNT/graphene and the influence of topology on electronic and optical properties. As a 2D composite, we consider a layer of parallel

SWCNTs located at the same distance from each other, enclosed between two graphene monolayers and covalently bonded to them. All in silico studies were carried out using the molecular dynamics method as well as AIREBO [18] and SCC DFTB 2 methods [19, 20]. As noted above, 2D films of CNT/graphene can mean different materials. On the one hand, we mean a graphene monolayer with an individual single-walled tube located on or under it. Herewith, there is a substrate of Si or SiO_2 layer [16]. On the other hand, most often a film means a monolayer of graphene with individual SWCNTs between them [1, 10, 21, 22]. Finally, there are rarely synthesized films where CNTs are not simply bonded to graphene but are interfaced with it. In our work, we will focus on the most common case of film synthesis when SWCNTs are located between two graphene monolayers at equal distances in parallel to each other and covalently bonded to graphene with the formation of sp^3 states, as shown in Refs. [23, 24]. At the same time, the mechanism of covalent binding of tubes and graphene in the composition of such films was not previously considered in scientific works as well as the thermodynamic stability of such hybrid compounds was not previously estimated. Also there are no data of the structural type and chirality of nanotubes in the 2D composite material. In this chapter, all these questions will be considered for the first time.

3.2 "Method of Magnifying Glass" for Constructing Atomic Models of New Composite Nanomaterial CNT/Graphene Supercells with Covalent Bonds

A new unique method for constructing atomistic models of new composite supercells has been developed. This technique was called the "Method of magnifying glass." This name completely reflects the essence of the method: Initially it is considered a large fragment of the composite by one of the corresponding empirical methods; then the analysis focuses on the small middle part of a large fragment with application of semi-empirical and non-empirical methods.

Let us consider an example of the "Method of magnifying glass" application for building an atomic supercell of CNT/graphene monolayer 2D composite that represents two monolayers of graphene with covalently bonded nanotubes between them.

To build topological models, we used SWCNTs with a diameter of ~0.8 nm ÷ 1.6 nm. As it is known, these SWCNTs are most often synthesized in macroscopic volumes. Acrylic SWCNTs were chosen since these tubes are suitable for the formation of covalent bonds with graphene. Tristán-López et al. [23] predicted that to form sp^3 states and covalent bonds between CNT and graphene (see Fig. 3.1d), the axis of the tube should be parallel to the larger diagonal of hexagonal graphene. Such contact is shown in Fig. 3.1a where the tube (12,0) is bonded to two graphene fragments. Covalent bonds are marked by orange. In this case, the edges of the tubes and graphene involved in the formation of covalent bonds are parallel to each other and form a rectangle. The common view of the monolayer composite fragment is shown in Fig. 3.1b. This is a fragment of five tubes between the graphene monolayer. However, the models shown in Fig. 3.1 were derived just in the zero approximation, since the true atomic structure of the composite is not known in advance. To find it, the following algorithm including several steps was applied.

3.2.1 First Step

In the first step, an extended fragment is constructed with tubes (n, 0) between two monolayer graphene sheets, as shown in Fig. 3.1b. The distance between the tubes is given by the number of hexagon in graphene between the ribs participating in the formation of covalent bonds. In Fig. 3.1b, the step is nine hexagons (we will denote it by 9H). The edges of graphene layers and tubes are passivated. The model was built using the original open-source KVAZAR (http://nanokvazar.ru), which was previously successfully used to study topological dielectrics [25], to model the process of graphene oxidation [26] and interaction of carbon structures with biomacromolecules [27]. The constructed fragment may contain tens of thousands of atoms (depending on the diameter of the tube, the position of the tubes, and the number of graphene layers).

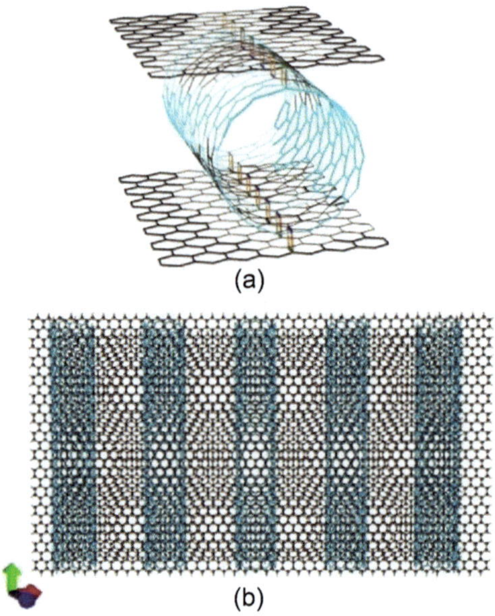

Figure 3.1 The view of the monolayer 2D composite atomistic model in the initial approximation (the tubes are marked with a blue color, graphene monolayer with the black one): (a) a fragment with a demonstration of the covalent bonds between the tube and graphene; (b) view from above (fragment with five tubes).

Therefore, in the first approximation, the search for an optimal atomic structure is performed by molecular dynamics using one of the known force fields. In our work, we used the AIREBO force field [18], which was successfully used for carbon nanostructures containing hydrogen atoms. The optimal structure was found by minimizing the total energy along the coordinates of all atoms of the 2D composite fragment.

3.2.2 Second Step

In the second step, the inner part is "cut out" from the optimized fragment, as shown in Fig. 3.2. This part should necessarily contain one middle tube and two on the sides. This creates an expanded supercell, which is two–three times less than the initial fragment in the first step. The expanded supercell is periodic only in the

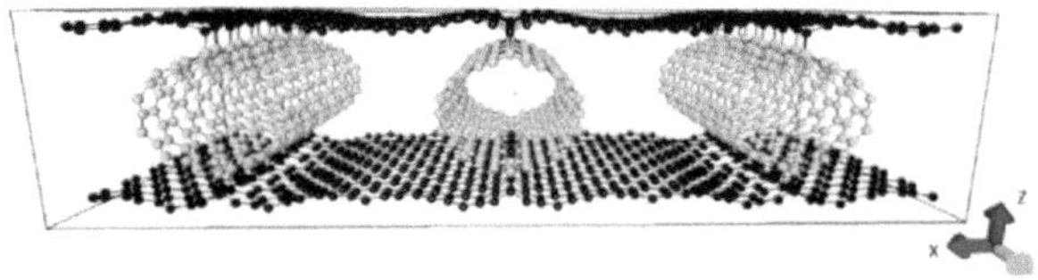

Figure 3.2 The expanded atomic cell of a monolayer composite after the structure's optimization by the empirical method.

x and y directions. Analysis of the atomic grid after optimization by the AIREBO method (Fig. 3.2) shows that the covalent bonds between CNT and graphene did not break and the bond length was equaled to 0.16 nm. The graphene became curvilinear and the tubes slightly deformed. In this step, the expanded cell of the composite is re-optimized with periodic boundary conditions by the molecular dynamics method using one of the known force fields. To refine the coordinates of the atoms, we used the AIREBO method again. Re-optimization was carried out with double optimization, since the dimensions of the box were optimized and for each box, the atomic grid was optimized by varying all the coordinates.

3.2.3 Third Step

In the next step, a new cell is cut from the optimized expanded cell of the composite, as shown in Fig. 3.3. This cell is also expanded, but it contains only the rings of three tubes, i.e., it narrows along the y-axis. This supercell is needed to obtain the atomic structure in the second approximation by one of the quantum methods. At this step, we applied the self-consistent charge density functional tight binding (SCC DFTB) 2 [19, 20] in the software DFTB+ (www.dftb-plus.info). The search for the optimal atomic structure of the constructed cell was carried out by the double optimization of a periodic box and atoms of its structure. This implied a step-by-step change in the lengths of the translational vectors L_x, L_y followed with the optimization of the supercell atomic mesh geometry. As a result, a matrix of minimum energies was obtained for each set L_x, L_y. From this matrix, the global energy minimum was determined. The result of double optimization is shown in Fig. 3.3. The topology of the composite has changed, the tube has become more noticeably

Figure 3.3 The expanded atomic cell of a monolayer composite after the structure's optimization by the quantum method.

deformed along the *z*-axis, and graphene layers have taken the form of regular waves. Covalent bonds between the CNT and graphene layers did not break, and the bond length slightly increased to 0.161 nm. This stage of searching for a true energetically favorable atomic structure is the next step in approaching the central fragment of the composite that will become a supercell of the composite. This can be compared with the successive enlargement of a small detail by a magnifying glass.

3.2.4 Fourth Step

In the last step, our virtual loop enlarges the middle-most part of the composite's extended fragment created in the first stage. From the extended cell of the previous step, we cut out its middle part, which is the supercell of the 2D composite CNT (12,0)/graphene. It undergoes re-optimization, which again consists in optimization of the periodic box and the structure's coordinates in it. In this step, one of the quantum methods (in our case SCC DFTB 2) is again applied as well as one of the methods for finding the global minimum of energy. For this topological model, the final values of the translational vectors were $L_x = 2.206$ nm, $L_y = 0.429$ nm.

Within the "Method of magnifying glass," an array of mono- and bilayer CNT/graphene composite models was built. Supercells of different models are shown in Fig. 3.4. Figure 3.4a presents models of a monolayer composite with tubes (10,0) and (12,0) with a distance between tubes of 10 and 12 hexagons. Exactly the

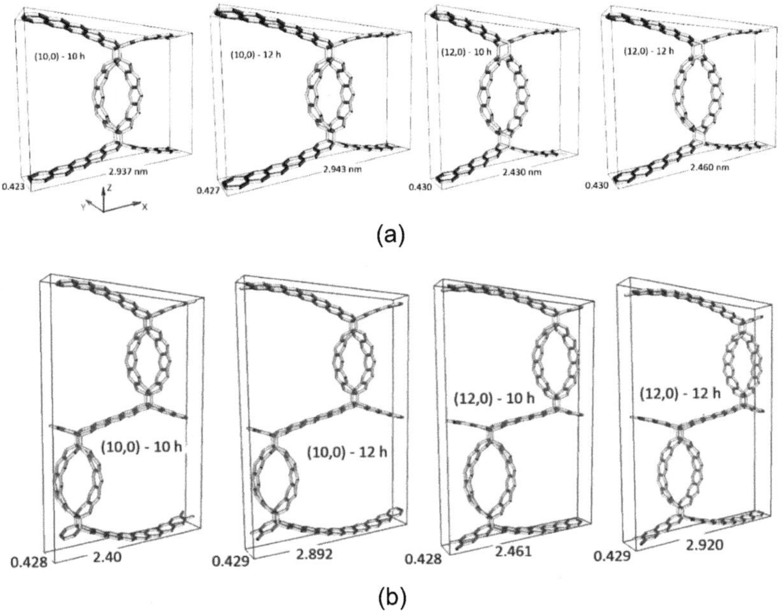

Figure 3.4 Supercells of mono- (a) and bilayer (b) composite films.

same models for the bilayer version of the composite are shown in Fig. 3.4b. In general, models of a composite with an SWCNT of 0.8 nm ÷ 1.6 nm with a distance of 7–15 hexagonals between the tubes were built. The thermodynamic stability of the obtained atomistic models of the composite was verified by the heat of formation, ΔH_f, of the composite from graphene and nanotubes. Analysis shows that tubes (n, 0) with only an even number n can form thermodynamically stable structures. For composites with odd n, the heat of formation is always positive and oscillates around 0.4 kcal/mol per atom (not less than 0.365 kcal/mol per atom). This follows from the geometry of the tube's atomic mesh, which does not allow creation of the stable covalent bond with graphene, as a result of which the framework of the tube and graphene is strongly deformed (which causes the positive heat of formation) or a covalent bond is not formed at all. Let us note that tubes with an even number n form a stable covalent bond with graphene due to the symmetry of the tube framework in relation to the plane containing a rectangle with covalent bonds of CNT graphene (Fig. 3.1a). For all the tubes (10,0), (12,0), (14,0), (16,0), (18,0), and (20,0), thermodynamically stable are the topological models of a monolayer composite with 9–13 hexagons between the tubes since the process of their formation is exothermic ($\Delta H_f < -0.1$ kcal/mol per atom). A small number of thermodynamically stable topological models were found for the bilayer composite as compared to the monolayer composite. According to our estimations, the formation of a bilayer composite with tubes (10,0)/(12,0) with a distance between tubes of 10/12/14/16 hexagons for tubes (10,0) and 10/12 hexagons for tubes (12,0) will be most energetically favorable. Its formation is exothermic with the heat of formation $H_f < -0.1$ kcal/mol per atom.

Mono- and bilayer composites have the same length of the covalent bond between CNT and graphene equaled to ~0.16 nm. The compression ratio of all tubes is almost the same and corresponds to the golden section of 1.618. In the bilayer composite, the translation vector L_y along the y-axis (along the tube axis) for all models is 0.43 nm, and the vector L_x along the x direction (perpendicular to the tubes) varies depending on the diameter of the tubes and the number of hexagons between them. For CNTs (10,0)/(12,0), L_x = 2.40/2.46 nm at a distance of 10 hexagons and L_x = 2.89/2.92 nm at a distance of 12 hexagons; for CNTs (10,0), L_x = 3.43 nm at a distance of 14 hexagons and L_x = 3.85 nm at a distance of 16 hexagons. The common view of monolayer and bilayer composites is shown in Fig. 3.5. Table 3.1 presents translation vectors of composites for distances between tubes of 10, 12, and 14 hexagons. Also it shows the distances between the tubes r_{t-t} and the parameter a/b (the ratio of the maximum tube dimensions along the z- and y-axes). The length of the covalent bond between the CNT and graphene in all samples is 0.161–0.1662 nm.

Table 3.1 shows that the compression ratio for all tubes is almost the same and corresponds to the golden section of 1.618. The periods of the composite lattice along both x- and y-axes practically do not change with an increasing diameter of the tube. It allows to suggest that the electronic and optical properties will be likely determined not by the tube but by the number of hexagons between them. Further, we investigate the patterns of the band structure and optical conductivity of the composite.

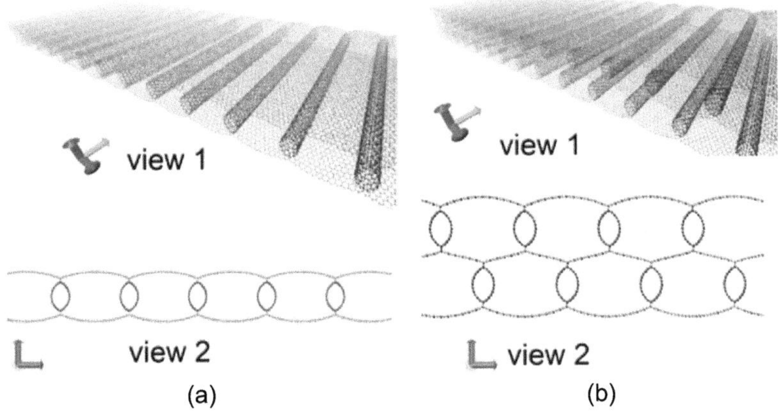

Figure 3.5 Atomistic models of mono- (a) and bilayer (b) CNT/graphene composites.

Table 3.1 Translation vectors, distances between the tubes, and the ratio of the maximum tube dimensions along the *z*- and *y*-axes for mono- and bilayer CNT/graphene composites

Parameters		(10,0)	(12,0)	(14,0)	(16,0)	(18,0)	(20,0)
10 hexagons	L_x, nm	2.207	2.206	2.207	2.203	2.20	2.20
	L_y, nm	0.427	0.429	0.428	0.429	0.42	0.430
	r_{t-t}, nm	1.593	1.549	1.326	1.234	1.086	0.929
	a/b	1.634	1.634	1.631	1.684	1.676	1.595
12 hexagons	L_x, nm	2.700	2.674	2.718	2.687	2.69	2.70
	L_y, nm	0.428	0.429	0.427	0.428	0.43	0.430
	r_{t-t}, nm	2.085	1.929	1.845	2.207	1.581	1.434
	a/b	1.684	1.662	1.661	1.610	1.688	1.604
14 hexagons	L_x, nm	3.220	3.240	3.218	3.218	3.186	0.318
	L_y, nm	0.428	0.428	0.427	0.428	0.430	0.43
	r_{t-t}, eV	2.596	2.454	2.344	2.219	2.034	1.92
	a/b	1.643	1.645	1.635	1.612	1.579	1.603

3.3 Electronic Properties of CNT/Graphene Composite

The density of electronic states (DOS) and the band structure of mono- and bilayer composites were calculated by SCC DFTB 2. For comparison, Fig. 3.6 shows diagrams of band structures for single-type models with one- and two-tube layers. This model is built on the base of the tubes (10,0) at a distance of 10 hexagons. The Brillouin zone (BZ) for all topological models of composites represents a rectangle since the supercell is periodic in only two directions. As it is known, the energy bands of graphene cross at K points—corner points in hexagons of the first Brillouin zone. The zigzag tubes, on the contrary, in most cases refer to semiconductor type of electronic conductivity. Figure 3.6a shows the first Brillouin zone and its highly symmetrical points. Also it shows that the direction of these points bypasses, starting with point M. In general, the trajectory is written in the form M-Γ-J-K-Γ. For tubes (12,0) and (18,0), valence and conduction bands intersect at the center of the BZ—point Γ. The composite with a semiconductor tube has a gap in the band structure between the conduction and valence bands. All the models of a monolayer composite with semiconductor tubes demonstrate a gap between the valence and conduction bands in several tenths of an electron volt. Herewith, the bilayer composite has no gap for all types of topological models. Figure 3.6b shows a gap VZ–CZ near the point Γ for a monolayer composite on the base of tubes (10,0), and in Fig. 3.6c the gap disappears when the composite is bilayer. The Fermi level is marked by a dotted line and does not change.

To find the difference in the electronic properties of mono- and bilayer composites, DOS spectra were calculated and analyzed. Figure 3.7 shows the DOS for the same topological models. It is seen that semiconductor tubes demonstrate the gap and the DOS becomes zero at the Fermi level. However, the bilayer structure does not show a gap. It is clearly seen that the electronic properties of a monolayer composite are determined only by the properties of CNTs.

Let us consider the regularities of the band structure formation for a monolayer composite in more detail. It should be noted that composites with tubes (10,0) cannot be attributed to this or that type of electronic conduction, since according to our studies, it is determined the distance between tubes, i.e., by the number of hexagons between them. At a distance of 9 hexagons, the valence band and the conduction band cross in the middle between the points Γ and J. As the step increases, a small gap appears and

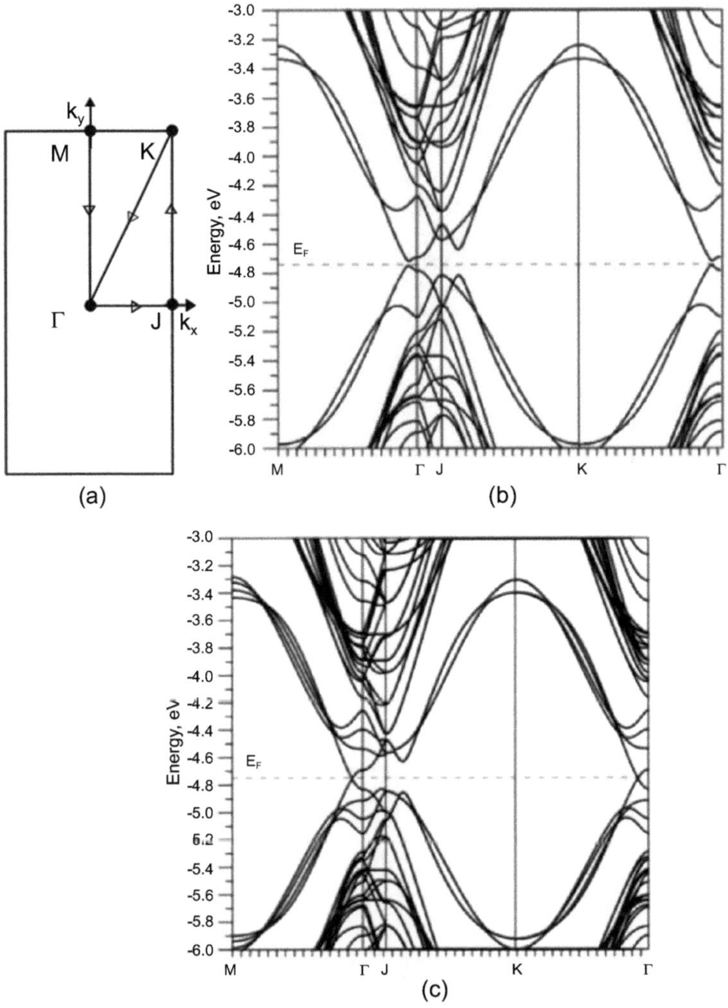

Figure 3.6 Brillouin zone (a), zone structure of monolayer (b), and bilayer (c) composite on the base of tubes (10,0) with distance of 10 hexagons.

reaches a maximum at 11 hexagons and then slightly decreases by several hundredths of an electron volt. Figure 3.8 shows the band structure near the Fermi level (it is marked by a red line) for a composite (10,0)-11 hexagons with a maximum gap of 0.1 eV for a given type of composite. We believe that the instability of

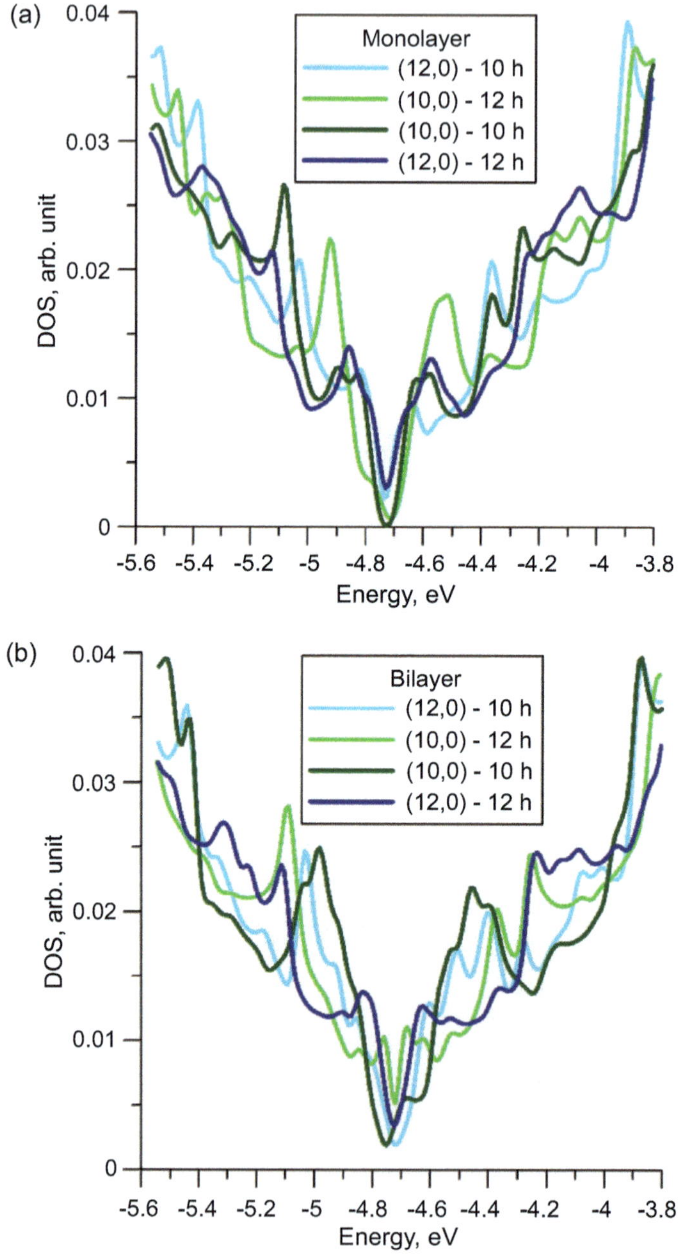

Figure 3.7 DOS for mono- and bilayer composites.

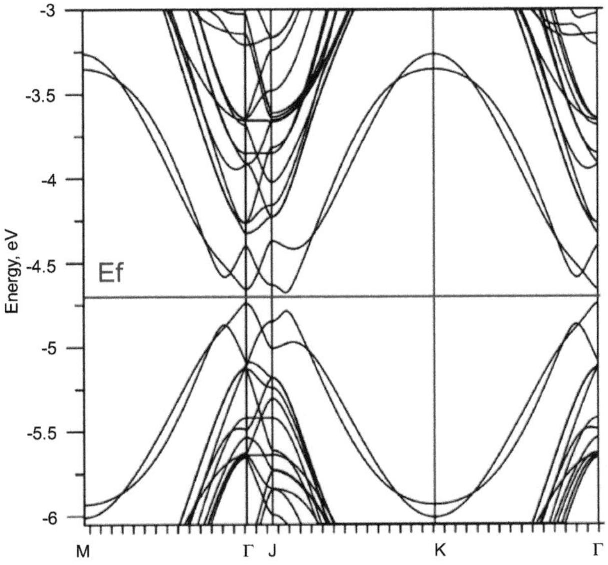

Figure 3.8 Zone structure of monolayer composite on the base of tubes (10,0) with a distance between tubes of 11 hexagons.

the electronic conductivity type can be explained by "competition" between the properties of a semiconductor tube (10,0) and the properties of corrugated graphene. The calculation of the isolated corrugated graphene band structure revealed a metallic type of conductivity with intersection of the zones to the left of the point Γ. It is observed for all distances between the tubes. The appearance of the tubes opens the gap, as shown in Fig. 3.8.

All composites with tubes (12,0) and (18,0) have a metallic type of conductivity; the conduction and valence bands intersect in the middle between the points Γ and J, as shown in Fig. 3.9a,b for the models (12,0)-13 hexagons and (18,0)-12 hexagons. It is interesting that for tubes (12,0) to the left of the point Γ and to the right of J, the bands practically converge but do not intersect, although the gap is insignificant and amounts to only ~0.03 eV. A similar picture is observed for tubes (18,0). To the left of the point Γ, the zones practically converge. The metallicity of conductivity for all configurations is explained by the metallicity of the tube itself and graphene.

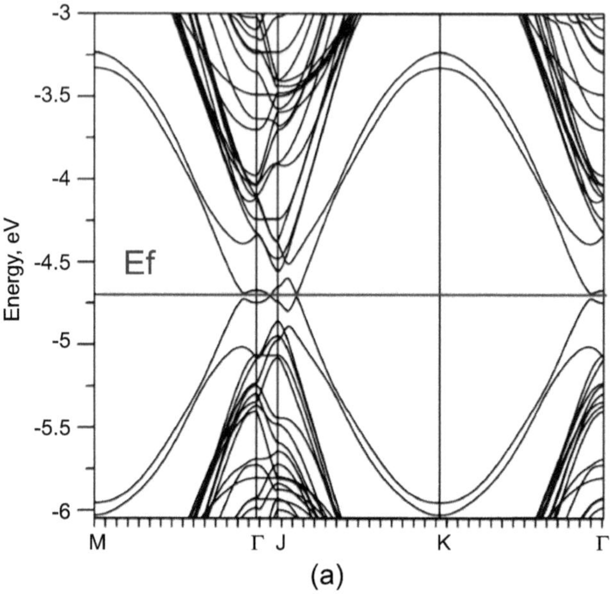

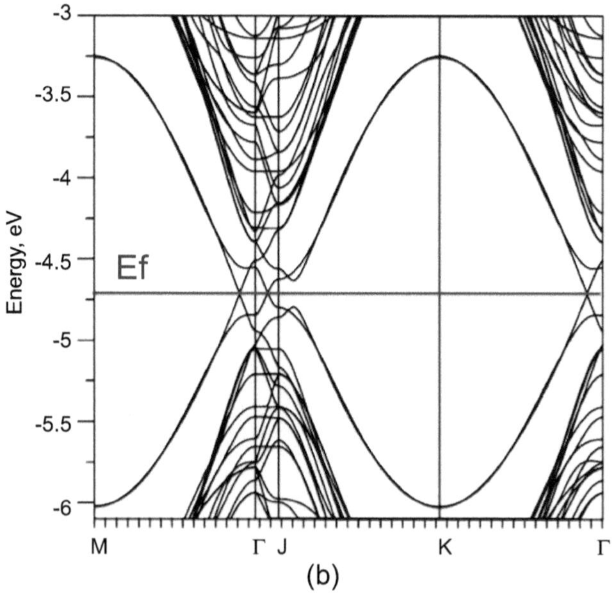

Figure 3.9 Zone structures of monolayer composites models: (a) (12,0)-13 hexagons; (b) (18,0)-12 hexagons.

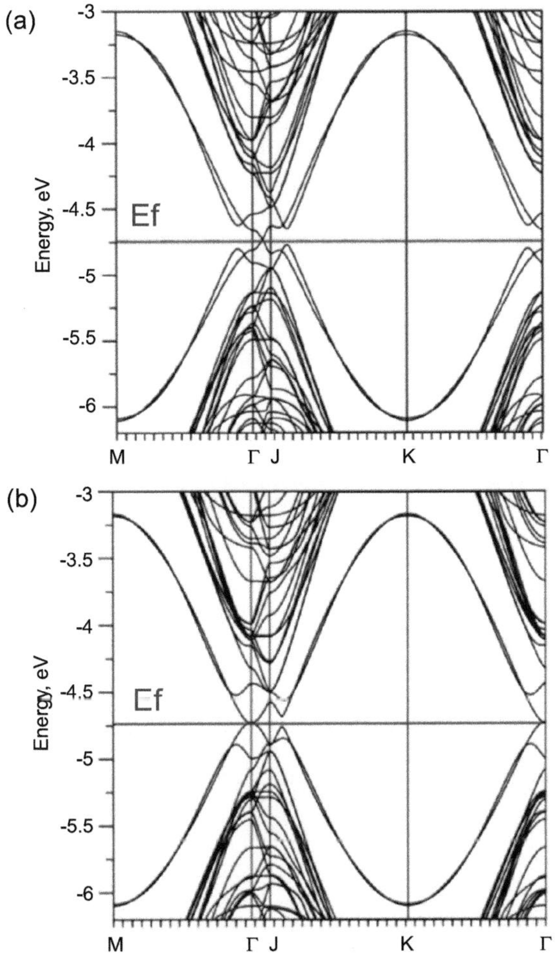

Figure 3.10 Band structures of monolayer composite models: (a) (14,0)-13 hexagons; (b) (16,0)-13 hexagons.

Composites with semiconductor tubes (14,0) and (16,0) demonstrate properties similar to the case of a tube (10,0). The corresponding band structures are shown in Fig. 3.10. When the distance between tubes is changed, the gap opens or the zones intersect. The gap is always small: 0.04–0.06 eV. For a composite with tubes (14,0) at a distance between tubes of 13 hexagons, the gap closes and the zones intersect at the same point as in the case of the tube (12,0)—in the middle between the points Γ and J. For a composite with tubes (16,0) at a distance of 13 hexagons, the zones intersect exactly at the point Γ.

Thus, based on the analysis of the calculated band structures, it can be concluded that CNT/graphene composites should be attributed to semimetals, since most of the considered topological models show a band gap of less than 0.06 eV. The only narrow-gap semiconductor is, apparently, the composite with tubes (10,0) and a distance between them of 10 or more hexagons. The gap is 0.08–0.1 eV. The Fermi level for all topological variations of the composite is $-4.73 \div -4.71$ eV.

3.4 Electrophysical Parameters of CNT/Graphene Composite

To estimate the prospects of the new composite, we investigated its electrical properties. The transmission functions and conductivity of the composite's various topological models were calculated. The study was carried out by the non-equilibrium Green's function method (NEGF) [28]. The conductivity of the object at low voltages can be determined by the transmission function:

$$G = 2\frac{e^2}{h}\int T(E)F_{\mathrm{T}}(E)dE \tag{3.1}$$

where

$$F_{\mathrm{T}}(E) = \frac{1}{4K_{\mathrm{B}}T}\sec h^2\left(\frac{E-\mu}{K_{\mathrm{B}}T}\right) \tag{3.2}$$

where μ is the Fermi energy of contacts, K_{B} is Boltzmann constant, T is temperature, e is electron charge, h is Planck's constant, and $T(E)$ is the electron transmission function. In our calculations, we used the method SCC DFTB 2.

First of all we calculated the conductivity functions of monolayer composites with different CNTs and different distances between them. Also graphs of conductivity for ideal nanotubes $(n, 0)$ at $n = 10, 12, 14, 16, 18$ were drawn. Figure 3.11a shows graphs of the transmission function $T(E)$ for a distance between tubes of

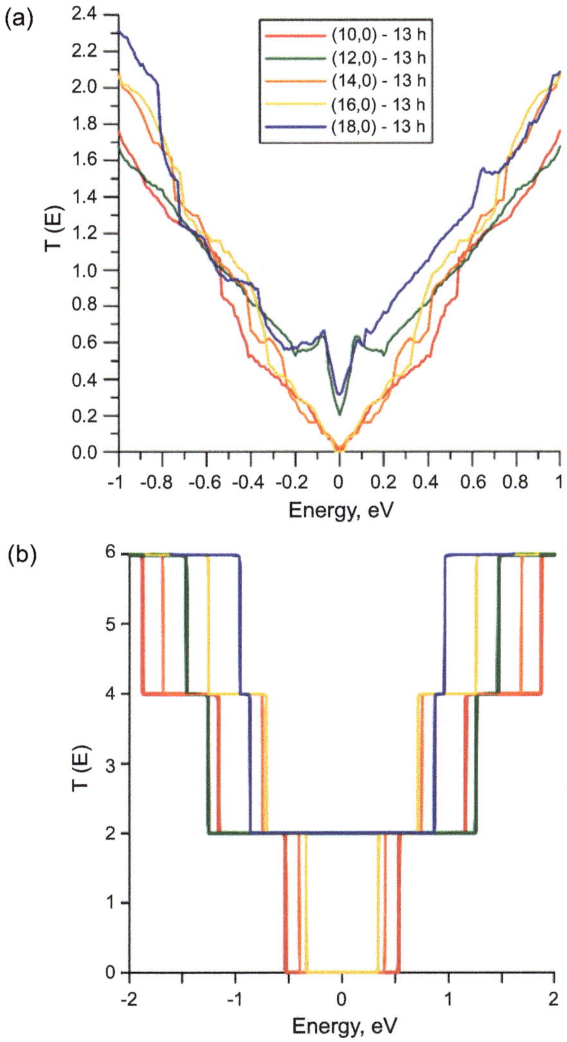

Figure 3.11 Transmission functions for a monolayer composite with different tubes (a) and for individual long tubes (b).

13 hexagons (units of measure $G_0 = e^2/h$). This figure vividly confirms the decisive effect of nanotubes on conductive properties of the composite. As it is well known, tubes of semi-conductive type (10,0), (14,0), and (16,0) have a gap of ~1 eV in the spectrum $T(E)$,

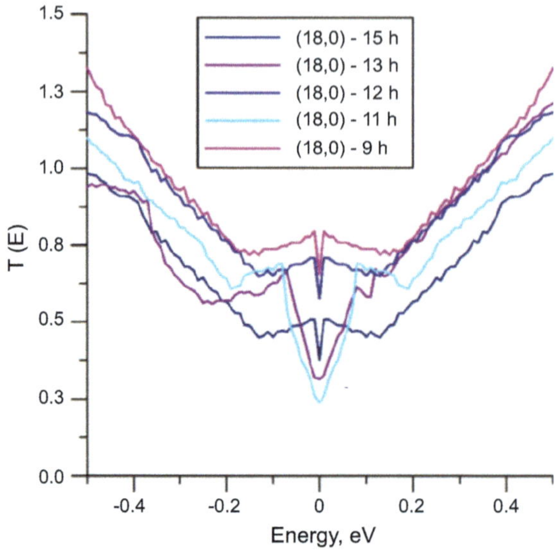

Figure 3.12 Transmission functions for a monolayer composite on the basis of tubes (18,0) at different distances.

while samples with tubes (12,0) and (18,0) do not have such a gap and their conductivity does not reach zero. However, composites with semiconductor tubes are characterized by a gap of 0.05–0.1 eV. It can be said that at room temperature and above, the conductivity G is nonzero. For comparison, Fig. 3.11b shows the transmission functions for tubes of the same types having a stepped view.

The role of the distance between the tubes of metallic conductivity is clarified in Fig. 3.12. It can be seen again that, similar to models of a composite with tubes (12,0), the maximum conductivity at the Fermi level is characterized by a model with the smallest distance equaled to 9 hexagons. Figure 3.12 clearly demonstrates that the electrical conductivity reaches its highest value at a distance between tubes of 9, 12, and 15 hexagons, i.e., at a distance of multiple three. It is important to note that at these distances, the graphs $T(E)$ qualitatively repeat each other.

The calculated transmission functions $T(E)$ of mono- and bilayer composites are shown in Fig. 3.13 (units of measure $G_0 = e^2/h$). Comparison of identical topological models of mono- and bilayer

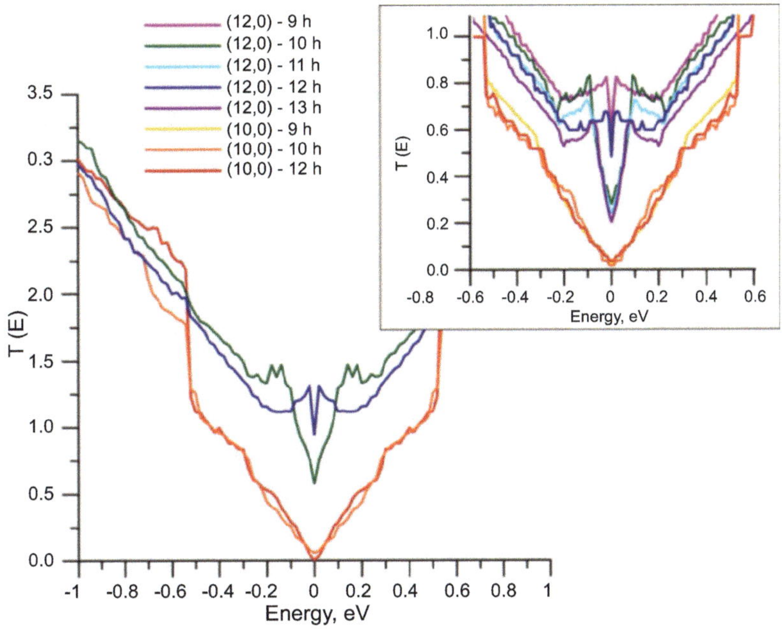

Figure 3.13 The transmission functions of the bilayer composite and monolayer (in frame) at various distances between the tubes and different chirality of the tubes.

composites shows that the graphs of $T(E)$ match qualitatively. It is clearly seen that the conductivity of the sample with CNT (10,0) at the Fermi level (0 eV) is practically zero. Herewith, the periodicity of the nanotube location and the number of layers do not play a role. On the contrary, in the composite with tubes (12,0), the connection between the value of the conductivity at the Fermi level and the distance between the tubes is clearly visible. Comparison of the same models corresponding to composites with different number of layers shows that the bilayer composite is characterized by a conductivity almost two times higher. This result is quite expected since the density of CNT packages in a bilayer composite practically doubles in comparison with a monolayer one. We also found that a monolayer composite with metallic CNTs exhibits the highest conductivity near the Fermi level ($T = 300$ K) at a distance between tubes of 9 and 12 hexagons, i.e., at a distance of multiple three.

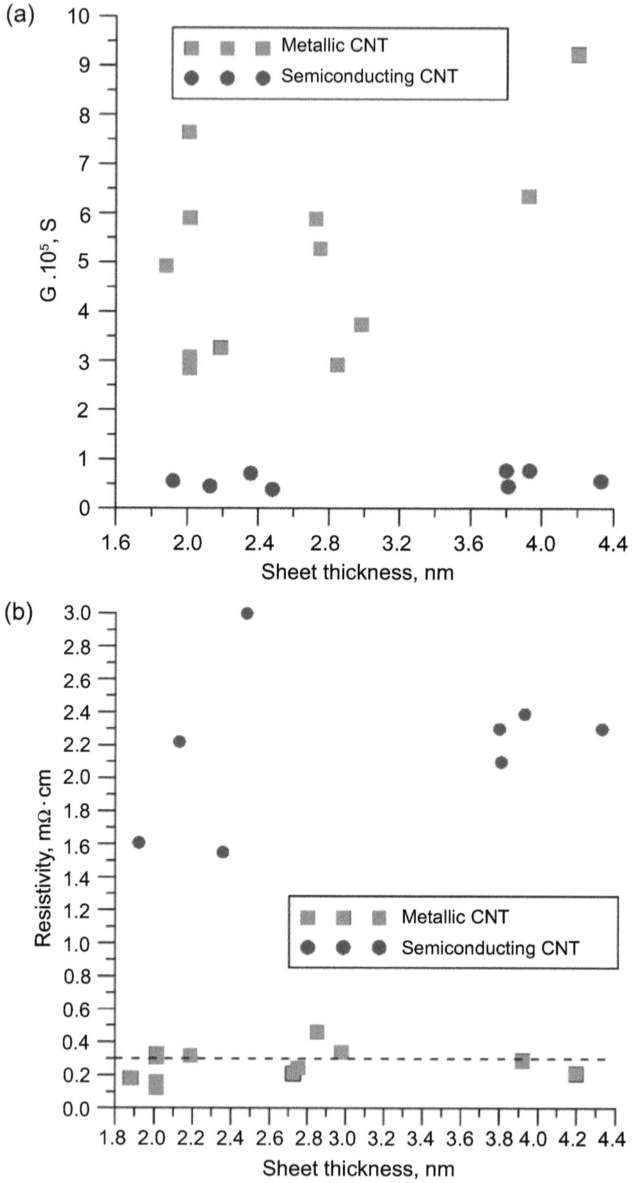

Figure 3.14 Dependence of (a) conductivity and (b) resistance on sheet thickness of the composite.

As it is known, the important characteristics of film materials used in physical and technical applications are surface and specific resistance. We calculated these characteristics depending on the thickness of the film taking account of its topology. Figure 3.14a shows the distribution diagram of the electrical conductivity for different films with different topological structures. The thickness was determined by the number of layers and the diameter of the tubes. The diagram shows a clear separation into films with semiconductive CNTs (blue balls) and films with metallic CNTs (green squares). The scatter in the conductivity of films with metallic tubes shows that bilayer films have a higher conductivity. The surface resistance was calculated as the value reverse to the conductivity G taken per unit area. To calculate the specific electrical resistance ρ, we took the film dimensions typical for the experimental samples 7.5 × 16 mm [23]. For both electrical parameters, there is no apparent dependence on the thickness. The maximum conductivity corresponds to the topological model of the bilayer film (12,0)-12 hexagons, and the largest specific electrical resistivity is observed in a monolayer film (16,0)-13 hexagons (Fig. 3.14b). The average value of the surface resistance of films with metal tubes was 20.94 kΩ/sq, and the average value of specific resistivity was 0.3 mΩ*cm. The smallest resistivity corresponds to the monolayer film (12.0)-9 hexagons, and regardless of film thickness, it lies in the interval (0.12 ÷ 0.46) mΩ*cm, which fits the resistance values of 0.3 mΩ*cm measured in Ref. [23].

3.5 Conclusion

The mechanism of covalent bonding between nanotubes and graphene in the new 2D composite CNT/graphene was studied. A new method was described for constructing topological models of hybrid CNT/graphene films corresponding to an energetically favorable configuration of a structure. This method consists in obtaining a refined topology of an unit cell of CNT/graphene composite with equilibrium geometrical parameters through successive magnification of the central fragment of the composite and finding its energetically favorable topology at each step of the magnification.

The fragment obtained in the last step becomes the elementary cell of the studied composite. The key procedure of the proposed technique that received the name "Method of magnifying glass" is the re-optimization of the structure, concluding in the relaxation scanning of the energy profile for the newly created cell. Re-optimization was performed with double optimization: optimization of periodic box sizes and optimization of the atomic mesh for each box by varying all the atom coordinates. The described procedure is also an original development of the authors. For the topological models of the 2D composite CNT/graphene obtained within this method, the following regularities were established for the first time:

- Composites with a covalent bond CNT/graphene can be thermodynamically stable only with tubes of the zigzag $(n, 0)$ type with an even number n;
- Stable topological configurations (heat of formation <0 kcal/mol) can be formed by tubes $(n, 0)$ with $n = 10$–20 (n even) at a distance between tubes of $8 \div 14$ hexagon;
- Composites with tubes (10,0), (16,0), (18,0), (20,0) at the distance of 6 hexagons and with tubes $n = 10$–20 (n even) at a distance of 16 hexagons are metastable morphological variations of the composite (heat of formation ~ 0 kcal/mol), which can be obtained during the synthesis of the composite, but they can exist only in the absence of external influences (increase in temperature, influence of electric fields, etc.)
- The most energetically favorable composite is the composite with a tube (12, 0) at a distance of 8, 9, and 10 hexagons (heat of formation <-0.8 kcal/mol per atom);
- For all thermodynamically stable composite models, the compression ratio of the tubes is approximately the same and corresponds to the golden section of 1.618);
- The periods of the composite lattice along the tube axes and perpendicular to them practically do not change with increasing diameter of the tube and at the constant distance between them.

For the analysis of electronic and electrical properties, we have chosen the most stable topological models (heat of formation <-0.2 kcal/mol per atom). This group of composites includes models with tubes (10,0), (12,0), (14,0), (16,0), and (18,0) at the intertube distances of 9, 10, 11, 12, and 13 hexagons.

Analysis of band structures shows that CNT/graphene composite should be attributed to semimetals since most topological models demonstrate band gap of less than 0.06 eV. The only one narrow-gap semiconductor is a composite with tubes (10,0) and a distance between them of 10 hexagons or more. The gap is 0.08–0.1 eV. The Fermi level for all topological variations of the composite is $-4.7 \div 4.71$ eV.

As a result, it can be concluded that a hybrid CNT/graphene film can be the elemental base of a new class of electronic devices, possibly nanoantennas. For this goal, these hybrid films have all the necessary qualities.

References

1. Jung, N., Kwon, S., Lee, D., Yoon, D.-M., Park, Y. M., Benayad, A., Choi, J.-Y., and Park, J. S. (2013). Synthesis of chemically bonded graphene/carbon nanotube composites and their application in large volumetric capacitance supercapacitors, *Adv. Mater.*, **25**, pp. 6854–6858.
2. Park, H., Choi, S., Lee, S., Hwang, G., Choi, N.-S., and Park, S. (2015). Novel design of silicon-based lithium-ion battery anode for highly stable cycling at elevated temperature, *J. Mater. Chem. A*, **3**, pp. 1325–1332.
3. Li, K., Eres, G., Howe, J., Chuang, Y.-J., Li, X., Gu, Zh., Zhang, L., Xie, S., and Pan, Zh. (2013). Self-assembly of graphene on carbon nanotube surfaces, *Sci. Rep.*, **3**, 2670.
4. Gan, X., Lv, R., Bai, J., Zhang, Z., Wei, J., Huang, Z.-H., Zhu, H., Kang, F., and Terrones, M. (2015). Efficient photovoltaic conversion of graphene–carbon nanotube hybrid films grown from solid precursors, *2D Mater.*, **2**, 034003.
5. Zhu, Y., Li, L., Zhang, Ch., Casillas, G., Sun, Zh., Yan, Zh., Ruan, G., Peng, Zh., Raji, A.-R. O., Kittrell, C., Hauge, R. H., and Tour, J. M. (2012). A seamless three-dimensional carbon nanotube graphene hybrid material, *Nat. Commun.*, **3**, 1225.

6. Xiao, F., Yang, Sh., Zhang, Zh., Liu, H., Xiao, J., Wan, L., Luo, J., Wang, Sh., and Liu, Y. (2015). Scalable synthesis of freestanding sandwich-structured graphene/polyaniline/graphene nanocomposite paper for flexible all-solid-state supercapacitor, *Sci. Rep.*, **5**, 9359.
7. Wang, W., Ozkan, M., and Ozkan, C. S. (2016). Ultrafast high energy supercapacitors based on pillared graphene nanostructures, *J. Mater. Chem. A*, **4**, pp. 3356–3361.
8. Higgins, D. C., Hoque, M. A., Hassan, F., Choi, J. Y., Kim, B., and Chen, Z. (2014). Oxygen reduction on graphene-carbon nanotube composites doped sequentially with nitrogen and sulfur, *ACS Catal.* **4(8)**, pp. 2734–2740.
9. Pham, D. T., Lee, T. H., Hoa, D., Yao, F., Ghosh, A., Le, V. T., Kim, T. H., Li, B., Chang, J., and Lee, Y. H. (2015). Carbon nanotube-bridged graphene 3D building blocks for ultrafast compact supercapacitors, *ACS Nano*, **9(2)**, pp. 2018–2027.
10. Zhang, Z., Wang, L., Xiao, J., Xiao, F., and Wang, S. (2015). One-pot synthesis of three-dimensional graphene/carbon nanotube/SnO_2 hybrid architectures with enhanced lithium storage properties, *ACS Appl. Mater. Interfaces*, **7(32)**, pp. 17963–17968.
11. Liu, Y., Wang, F., Wang, X., Wang, X., Flahaut, E., Liu, X., Li, Y., Wang, Xinran, Xu, Y., Shi, Y., and Zhang, R. (2015). Planar carbon nanotube–graphene hybrid films for high-performance broadband photodetectors, *Nat. Commun.*, **6**, 8589.
12. Cheng, Y., Lu, S., Zhang, H., Varanasi, C.V., and Liu, J. (2012). Synergistic effects from graphene and carbon nanotubes enable flexible and Robust electrodes for highperformance supercapacitors, *Nano Lett.*, **12(8)**, pp. 4206–4211.
13. Koutsioukis, A., Georgakilas, V., Belessi, V., and Zboril, R. (2017). Highly Conductive Water–Based Polymer/Graphene Nanocomposites for Printed Electronics, *Chemistry*, **23(34)**, pp. 8268–8274.
14. Pan, Y., Hou, Z., Yi, W., Zhu, W., Zeng, F., and Liu, Y. N. (2015). Hierarchical hybrid film of MnO_2 nanoparticles/multi-walled fullerene nanotubesegraphene for highly selective sensing of hydrogen peroxide, *Talanta*, **141**, pp. 86–91.
15. Wu, S., Shi, E., Yang, Y., Xu, W., Li, X., and Cao, A. (2014). Direct fabrication of carbon nanotube-graphene hybrid films by a blown bubble method, *Nano Res.*, **8(5)**, pp. 1746–1754.

16. Bonaccorso, F., Colombo, L., Yu, G., Stoller, M., Tozzini, V., Ferrari, A. C., Ruoff, R. S., and Pellegrini, V. (2015). Graphene, related two-dimensional crystals, and hybrid systems for energy conversion and storage, *Science*, **347(6217)**, pp. 1246501-1–1246501-9.
17. Xiao, P., Wan, C., Gu, J., Liu, Z., Men, Y., Huang, Y., Zhu, L., and Chen, T. (2015). 2D Janus hybrid materials of polymer-grafted carbon nanotube/graphene oxide thin film as flexible, miniature electric carpet, *Adv. Mater.*, **25(16)**, pp. 2428–2435.
18. Stuart, S. J., Tutein, A. B., and Harrison, J. A. (2000). A reactive potential for hydrocarbons with intermolecular interactions, *J. Chem. Phys.*, **112**, pp. 6472–6486.
19. Elstner, M., Porezag, D., Jungnickel, G., Elsner, J., Haugk, M., Frauenheim, Th., Suhai, S., and Seifert, G. (1998). Self-consistent-charge density-functional tight-binding method for simulations of complex materials properties, *Phys. Rev. B*, **58**, pp. 7260–7268.
20. Elstner, M. and Seifert, G. (2014). Density functional tight binding, *Phil. Trans. R. Soc. A*, **372**, 20120483.
21. Cui, X., Lv, R., Sagar, R. U. R., Liu, C., and Zhang, Z. (2015). Reduced graphene oxide/carbon nanotube hybrid film as high performance negative electrode for supercapacitor, *Electrochim. Acta*, **169**, pp. 342–350.
22. Chen, S., Yeoh, W., Liu, Q., and Wang, G. (2012). Chemical-free synthesis of graphene–carbon nanotube hybrid materials for reversible lithium storage in lithium-ion batteries, *Carbon*, **50**, pp. 4557–4565.
23. Tristán-López, F., Morelos-Gómez, A., Vega-Díaz, S. M., García-Betancourt, M. L., Perea-López, N., Elías, A. L., Muramatsu, H., Cruz Silva, R., Tsuruoka, S., Kim, Y. A., Hayahsi, T., Kaneko, K., Endo, M., and Terrones, M. (2013). Large area films of alternating graphene–carbon nanotube layers processed inwater, *ACS Nano*, **7(12)**, pp. 10788–10798.
24. Lv, R., Cruz-Silva, E., and Terrones, M. (2014). Building complex hybrid carbon architectures by covalent interconnections: Graphene-nanotube hybrids and more, *ACS Nano*, **8(5)**, pp. 4061–4069.
25. Pincak, R., Smotlacha, J., and Osipov, V. A. (2015). Electronic states of zigzag graphene nanoribbons with edges reconstructed with topological defects, *Physica B: Condens. Matter*, **475**, pp. 61–65.
26. Shunaev, V. and Glukhova, O. E. (2016). Topology influence on the process of graphene functionalization by epoxy and hydroxyl groups, *J. Phys. Chem. C.*, **120(7)**, pp. 4145–4149.

27. Glukhova, O. E., Prytkova, T. R., and Savostyanov, G. V. (2016). Simulation of high density lipoprotein behavior on a few layer graphene undergoing non-uniform mechanical load, *J. Phys. Chem. B*, **120(15)**, pp. 3593–3600.

28. Marder, M. P. (2011). Condensed Matter Physics, 2nd ed. (Department of Physics, University of Texas, Austin, USA).

Chapter 4

Control of the Emission Properties of Pillared Graphene

Olga E. Glukhova[a,b] **and Dmitry Shmygin**[a]
[a] *Physical Department, Saratov State University, 83 Astrakhanskaya Street 410010, Saratov, Russia*
[b] *I. M. Sechenov First Moscow State Medical University, Laboratory of Biomedical Nanotechnology, Trubetskaya Street 8-2, Moscow, 119991, Russia*
glukhovaoe@info.sgu.ru

This chapter is devoted to the study of the emission properties of two-dimensional (2D) films of pillared-graphene-doped potassium. We considered some topological model of mono- and bilayer pillared graphene with tubes of different lengths but with the same diameter. The charge transfer, electronic structure, and regularities of the Fermi level shift for the investigated pillared graphene films are calculated.

4.1 Introduction

Carbon nanotubes (CNTs) and graphene have unique mechanical, electronic, thermal, and emission properties. They are used as the material for point electron sources, X-ray sources, and emitters in flexible displays. There are ways to improve the useful properties

2D and 3D Graphene Nanocomposites: Fundamentals, Design, and Devices
Edited by Olga E. Glukhova

ISBN 978-981-4800-41-9 (Hardcover), 978-0-429-20150-9 (eBook)
www.jennystanford.com

of carbon nanomaterials. For example, there are different ways to improve the emission properties of CNTs: ion beam irradiation of material surface [1], new composite materials construction, and alkali metals addition. The change in emission properties can be tracked from the change in the value of the work function. Let us dwell on the last method of improving the emission properties. In Ref. [2], the work function values for multiwalled carbon nanotubes (MWNTs) are given after the addition of lithium atoms to the system.

The decrease in the work function was $\sim$0.4 eV. It was shown in Ref. [3] that the addition of potassium atoms to double-walled CNTs led to a decrease in the electric field strength required for the onset of field emission, and also to an increase in the emission current density. In Ref. [4], the work function of CNTs decreased by 0.46 eV, when lithium atoms were added. The authors of Ref. [5] carried out a series of experiments on the effect of cesium and potassium atoms on the work function of an emitter created on the basis of multilayered CNTs, and also obtained current–voltage characteristics of field emission for various methods of depositing cesium/potassium atoms on the emitter [5]. The method of depositing alkali metal atoms on the emitter surface has a significant effect on the type of current–voltage characteristic of the field emission. The minimum value of the work function that was achieved for emitters from MWNTs was 3.1–3.2 eV for cesium deposition for 20–30 s, and 2.9 eV for the deposition of potassium atoms for 180 s.

In Ref. [6], the effect of addition of cesium atoms on the properties of a single CNT was investigated. A decrease in the work function from 4.8 to 3.7 eV was noted.

Similar studies for 2D and 3D samples of pillared graphene were not conducted in both experimental and theoretical methods. Since pillared graphene has a large number of pores and cavities suitable for filling, in a series of numerical experiments, the effects of alkali atoms (particularly, potassium) on the magnitude of its work function and the density of electronic states were considered.

The Fermi energy E_F [7] can be found from the following equation:

$$2\sum_{i=1} f\left(\varepsilon_i, E_F\right) = N \tag{4.1}$$

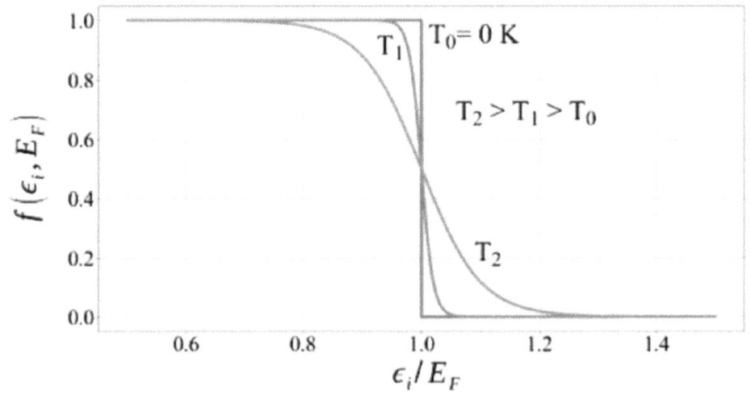

Figure 4.1 Common view of the Fermi–Dirac function for three different temperatures.

where f is the Fermi–Dirac function, ε_i is the electron energy in the state i, N is the total number of electrons in the system, and the factor 2 takes into account the presence of spin. The Fermi–Dirac function shows the probability of finding an electron at a level with energy ε_i:

$$f\left(\varepsilon_\mathrm{i},\, E_\mathrm{F}\right) = \frac{1}{1 + \exp\left(\frac{\varepsilon_\mathrm{i} - E_\mathrm{F}}{k_\mathrm{B} T}\right)} \tag{4.2}$$

In this equation, k_B is the Boltzmann constant and T is the temperature at which the molecular system is investigated. Figure 4.1 schematically shows a graph of $f\left(\varepsilon_\mathrm{i},\, E_\mathrm{F}\right)$ for different temperatures.

As can be seen from Fig. 4.1, at absolute zero there is a sharp transition from filled states to free ones. With increasing temperature, the sharpness of the transition is smoothed, since there is a nonzero probability of finding an electron in a state with a higher energy. The value of the Fermi energy for the molecular structure at low temperatures and the absence of an external electric field show the maximum energy that an electron can have, which uniquely determines the work function of the electron from the material. The presence of a small temperature has a weak effect on this value.

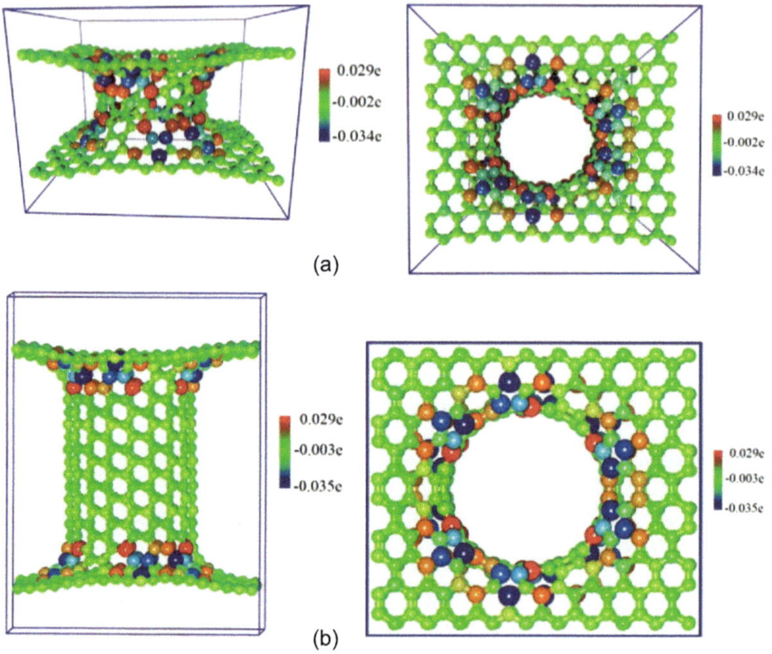

Figure 4.2 The charge distribution in the cells of a monolayered pillared graphene with CNT lengths (a) 0.6 nm and (b) 1.84 nm. Periodic boundaries are drawn.

4.2 Emission Properties of Pillared Graphene without Potassium Atoms

The charge density distribution over the particles of the molecular system for a monolayer composite is shown in Figs. 4.2 and 4.3.

Figure 4.2 shows that the charge is redistributed at the defect sites more actively than inside the defect-free part of the CNT or graphene sheet. Figure 4.3 shows the remaining patterns of charge distribution in monolayer composites in the absence of potassium atoms.

It can be seen from Figs. 4.2 and 4.3 that not only the charge concentrates in the places of defects, but its magnitude on the atoms in the defects does not depend significantly on the length of the CNTs in the composite.

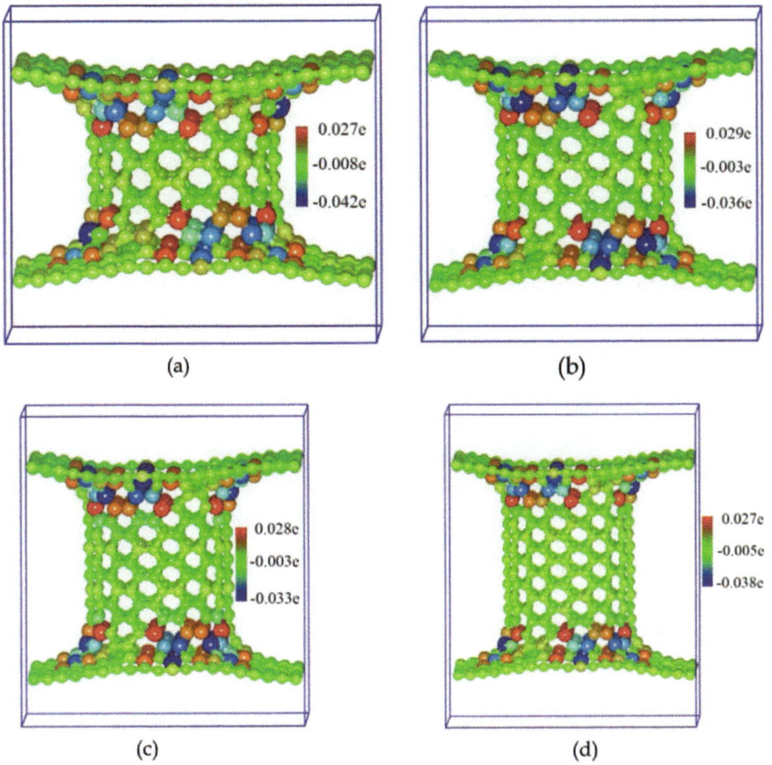

Figure 4.3 Charge redistribution for supercells of monolayered pillared graphene films with CNT lengths (a) 0.85 nm, (b) 1.1 nm, (c) 1.34 nm, and (d) 1.59 nm. Periodic boundaries are drawn.

The charge distribution was obtained during the optimization of the geometry of monolayered films by the conjugate gradient method with the usage of the electron density functional method with the charge self-consistency procedure (SCC-DFTB).

For bilayered pillared graphene films, the situation with the distribution of charges is completely analogous to the previous case—the charge distribution actively passes in the regions containing defects—at the junction of CNTs and graphene sheets (Fig. 4.4).

The procedure of calculating the charge distribution is still the same that was used earlier, but the optimization process was carried out using molecular dynamics with the AIREBO force field, since the

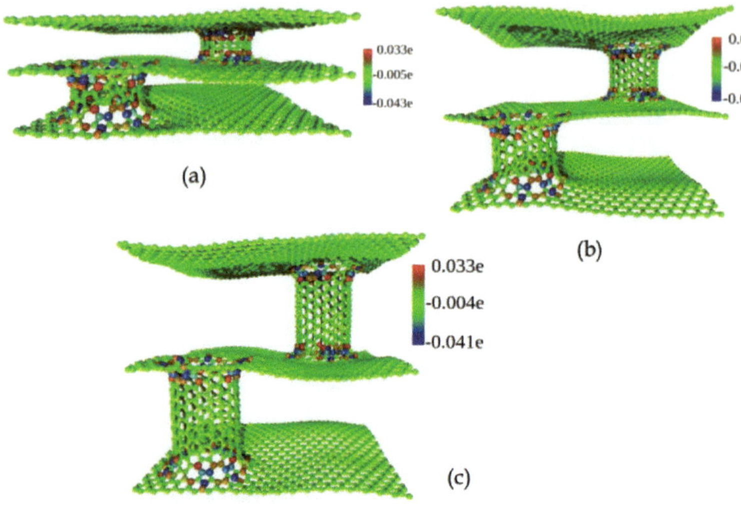

Figure 4.4 Charge distribution in the supercells of pillared graphene bilayer. Periodic boundaries are not shown.

energy calculation by SCC-DFTB of such large structures takes a very long time.

The solution to the large structures problem for a three-dimensional (3D) composite is similar. The charge distribution corresponds to a previously determined pattern (Fig. 4.5). The values of the Fermi energy of the composite films and 3D pillared graphene without potassium atoms are presented in Table 4.1.

The work function is slightly lower for bilayered films with respect to monolayered films (~0.129 eV). In the 3D composite, the work function is still slightly reduced (~0.022 eV).

The densities of electronic states for films and 3D composites without potassium atoms are presented in Figs. 4.6 and 4.7.

All the graphs above are normalized to the number of particles in a unit cell. According to the density of electronic states, the material has allowed states near the Fermi level. The graphs are similar to each other, since the general view of the structures is topologically the same. Nevertheless, the density of electronic states normalized to the number of particles near the Fermi energy differs for the models considered. For example, Fig. 4.7 shows that for a pillared

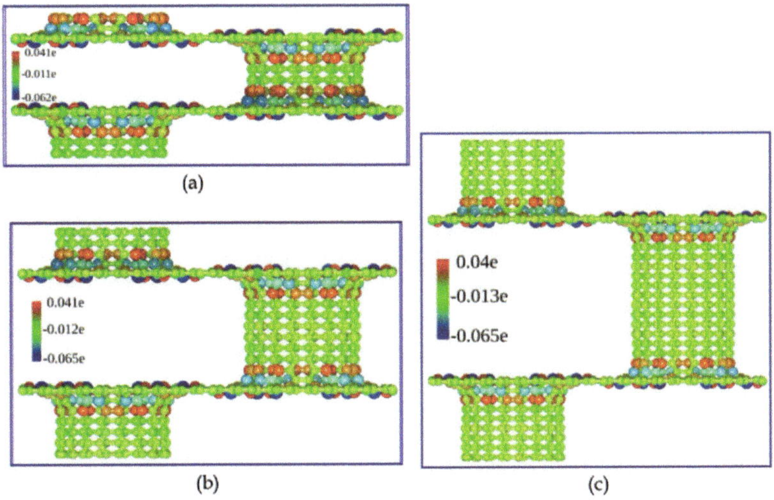

Figure 4.5 Charge distribution in a composite bilayer. The boundaries of a periodic cell are drawn.

Table 4.1 Work function for composite films and 3D pillared graphene for different lengths of CNTs

CNT length (nm)	Number of atoms in supercell	Work function (eV)
Composite monolayer		
0.60	400	4.616
0.85	436	4.666
1.10	472	4.624
1.34	508	4.631
1.59	544	4.646
1.84	580	4.626
Composite bilayer		
0.60	2400	4.505
1.10	2544	4.507
1.84	2760	4.506
3D composite		
0.60	1600	4.492
1.10	1744	4.482
1.84	1960	4.475

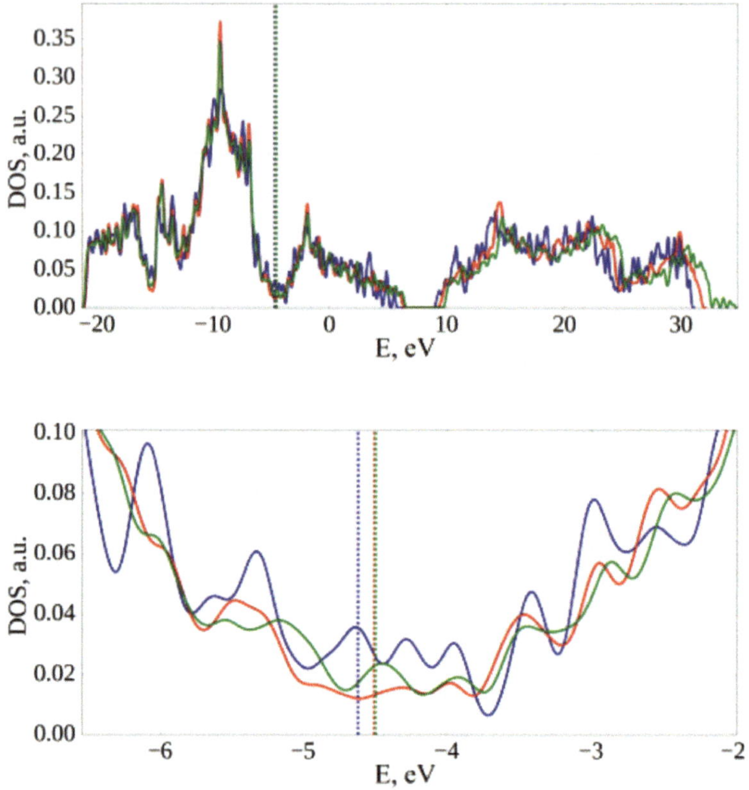

Figure 4.6 The density of electronic states for composite monolayer (blue), bilayer (red), and 3D structure (green) at a 0.6 nm length of CNTs.

graphene with a CNT length of 1.84 nm, the density of states is higher in a pure monolayer film, and less in a bilayered film. The same conclusion can be made from Fig. 4.6 for pillared graphene with a CNT length of 0.6 nm.

4.3 Emission Properties of Pillared Graphene at Different Mass Fraction of Potassium Atoms

One way to improve the emission characteristics of CNTs is to intercalate them with alkali metal atoms, such as lithium, cesium,

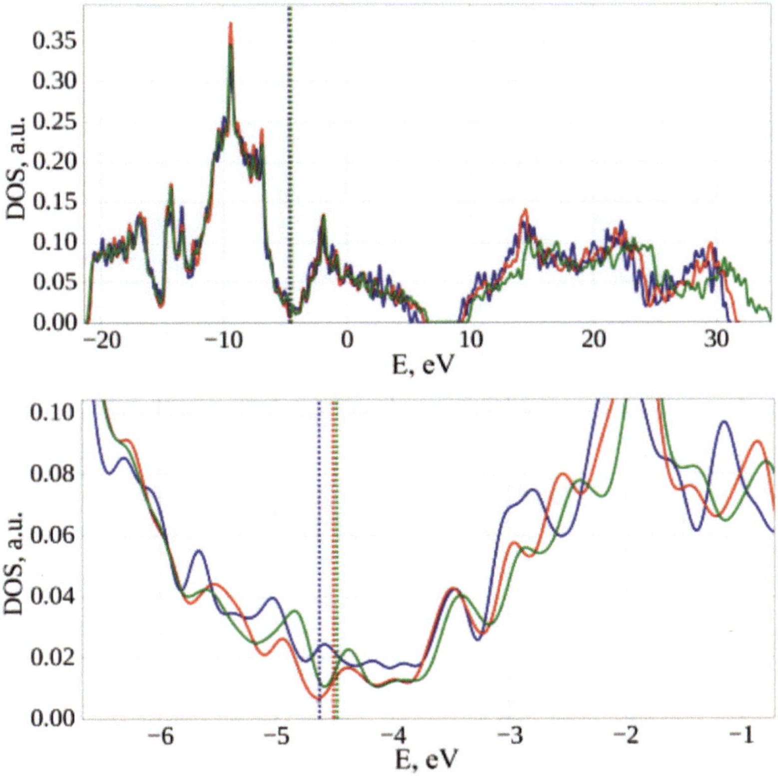

Figure 4.7 The density of electronic states for composite monolayer (blue), bilayer (red), and 3D structure (green) at a 1.04 nm length of CNTs.

and potassium. It is known that the presence of alkali metals in CNTs leads to a decrease in the work function [8]. It was shown in Ref. [2] that the work function of MWNTs decreased by ~0.4 eV when MWNTs were doped with lithium atoms. In Ref. [4], the work function is also reduced by ~0.46 eV. The intercalation of CNTs with potassium and cesium atoms leads to a substantial decrease in the work function, which was shown in Ref. [5] as a result of experiments. It was also noted that in practice, the results obtained depend on the time of deposition of cesium and potassium on the emitters under investigation on the basis of MWNTs (Fig. 4.8). It was noted that the work function value decreased by 1.6 eV and 1.8 eV for the deposition of cesium and potassium atoms, respectively. Such

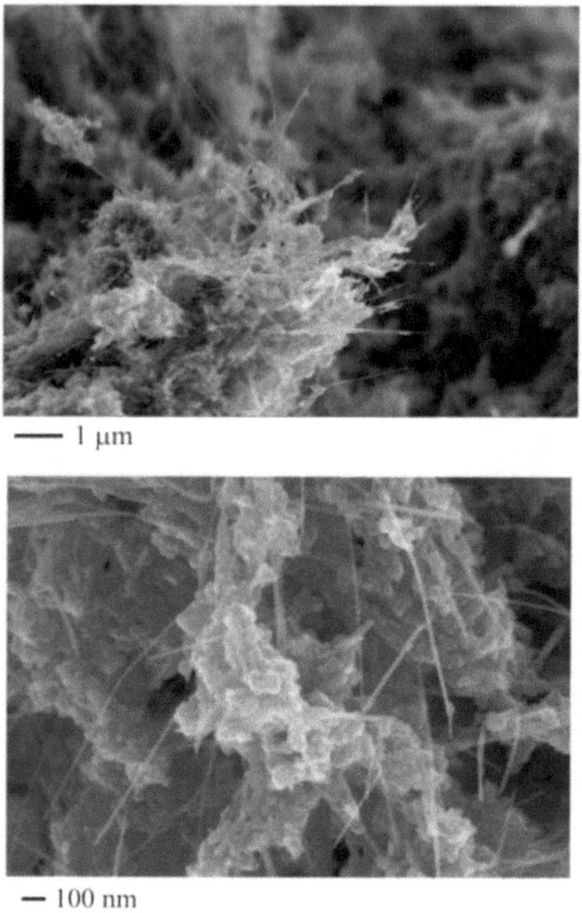

Figure 4.8 The part of the surface area of the emitter with multiwalled CNTs emerging from the cathode deposit, grown in an electric arc. Reprinted from Ref. [5], Copyright 2013, with permission from Elsevier.

a change in material properties leads to an increase in emission currents and a decrease in the threshold fields for low-voltage field emission [8].

To find the effect of potassium atoms on the emission characteristics of pillared graphene, a numerical experiment was carried out. First, the potassium atoms were added to the unit cells of a pillared graphene monolayer (Fig. 4.9).

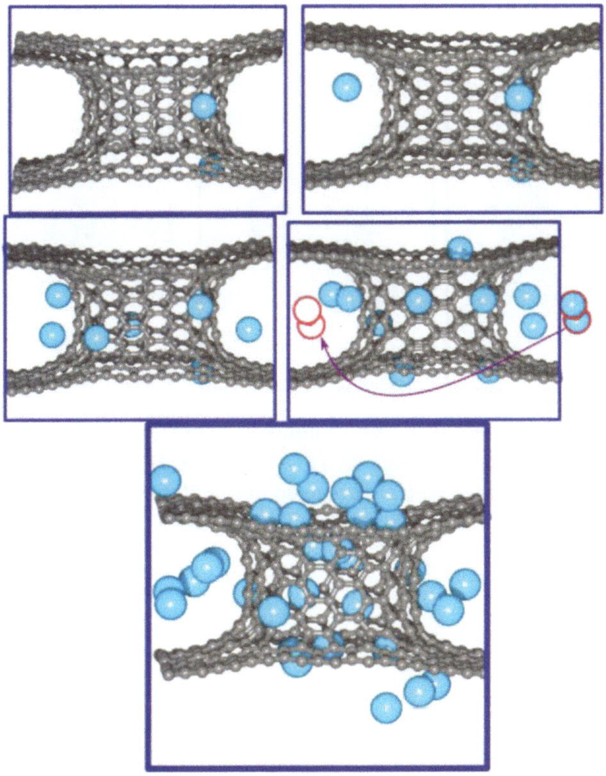

Figure 4.9 Elementary cells of a monolayer pillared graphene film where CNT has a length of 0.6 nm filled with potassium atoms (2, 4, 8, 14, 33 atoms). The arrow indicates the position of the atoms, which were used in calculations, taking into account the boundaries of the periodic cell. The cells are not periodic along the vertical direction.

Figure 4.10 shows several supercells of atomistic models of pillared graphene monolayer with the largest of the considered lengths of CNTs.

For monolayer films, the heat of formation was calculated with the addition of potassium atoms: the total energy of the optimized unit cell of the composite and *N* potassium atoms was subtracted from the resulting potential energy of the new system. The potential energy of a single atom (−2.319 eV) was calculated using the same method and the same parameters set (Fig. 4.11).

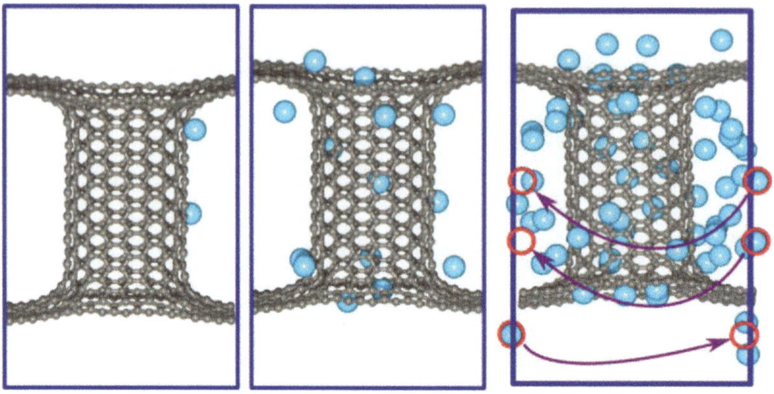

Figure 4.10 Elementary cells of monolayer pillared graphene with CNTs 1.84 nm long with potassium—2, 14, 61 particles. The arrows indicate the positions of the atoms, which were used in calculations, taking into account the boundaries of the periodic cell. The cells are not periodic along the vertical direction.

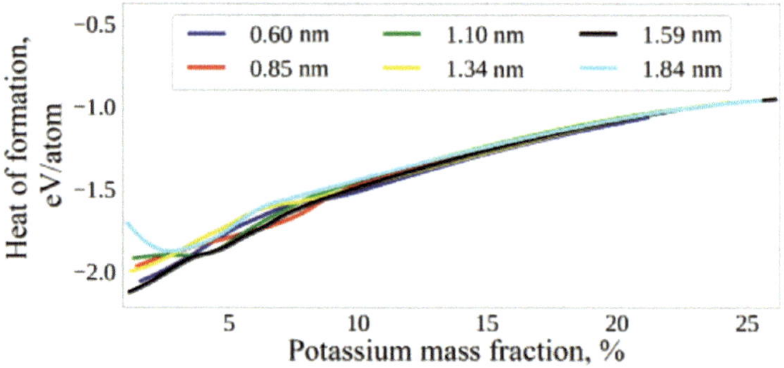

Figure 4.11 The heat of formation for a different number of potassium atoms in the system for all different lengths of CNTs in the system.

For all the situations considered, the heat of formation is negative and this shows stability of these configurations. However, the heat of formation increases with increasing the number of potassium atoms, which is due to repulsion between the atoms of the alkali metal. The average distance between the potassium particles and the carbon skeleton is 2.997–3.004 Å for all the situations considered. When the intercalated atoms and the carbon composite

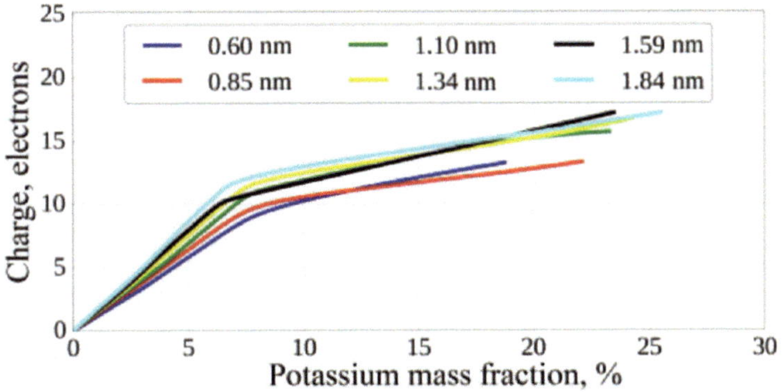

Figure 4.12 The value of the transferred charge from the potassium atoms to the carbon composite at different mass fractions of potassium in the system.

interact, the charge is transferred to the column graphene. The value of the transferred charge depends on the mass fraction of potassium and is shown in Fig. 4.12.

It can be seen from the graph that the increase in the value of the transferred charge is nonlinear. It will be legitimate to assert that from some value of the mass fraction of potassium atoms, saturation will occur in the system. This behavior is typical for models of all considered configurations of the composite. The parts of the graphs nearest to the zero mass fraction have linear form, because each new atom was added to the metal atoms located in system at the previous stages.

Figure 4.13 shows the plots of the values of the transferred charge normalized to the number of added alkali metal atoms.

It can be seen from Fig. 4.13 that with an increase in the number of potassium atoms from a certain moment, the participation of each particle in the process of charge transfer decreases due to the screening of the charge by the atoms near the carcass.

The charge transferred from potassium to pillared graphene is unevenly distributed in the carbon composite, concentrating on defects.

Also charge localization depends on the proximity of intercalated atoms to the pillared graphene carbon frame (Figs. 4.14 and 4.15).

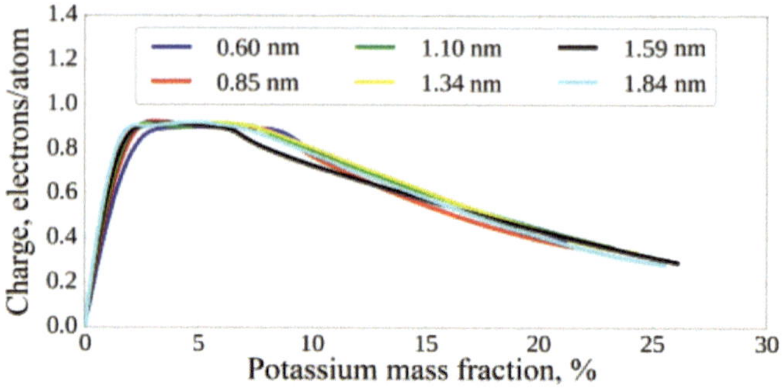

Figure 4.13 Dependence of the value of transferred charge from potassium atoms to a carbon composite at different mass fractions of potassium in the system, normalized to the number of potassium atoms.

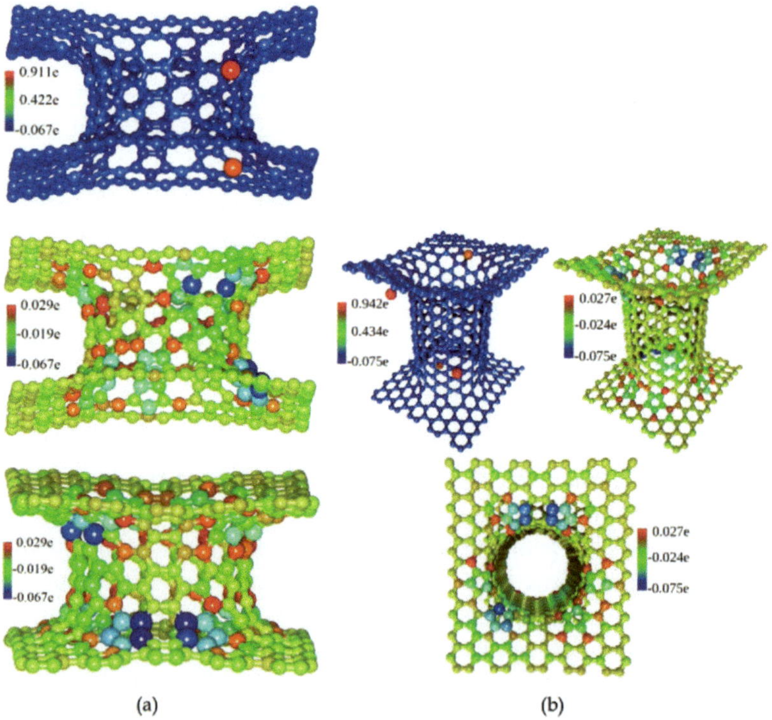

Figure 4.14 Charge distribution for two configurations of the composite: (a) length of CNT is 0.6 nm, two potassium atoms; (b) length of CNT is 1.84 nm, four potassium atoms. Periodic boundaries are not specified.

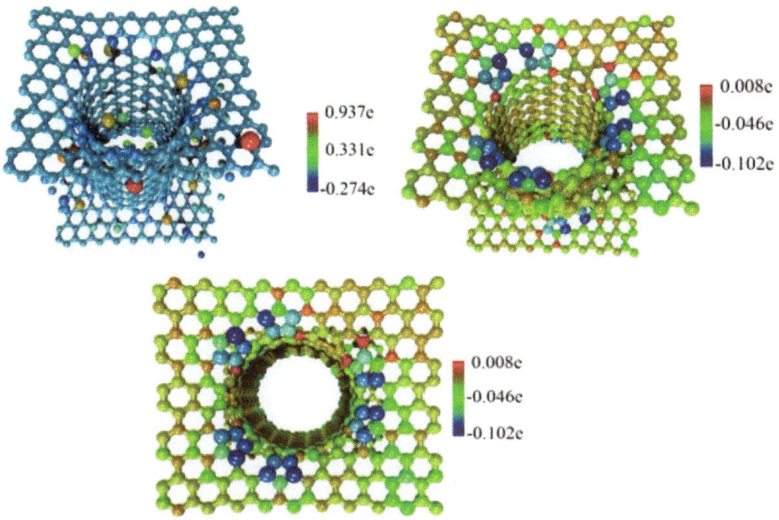

Figure 4.15 Charge distribution in a pillared graphene monolayer with a CNT length of 1.84 nm. The number of potassium atoms in the system is 61. Periodic boundaries are not indicated.

In addition, Fig. 4.15 shows the distribution of charges on atoms of the molecular system in the case of the maximum mass fraction of potassium atoms from the considered.

Intercalation of a monolayer film of pillared graphene by potassium atoms led to a significant decrease in the Fermi energy and, accordingly, the work function of composite. The difference

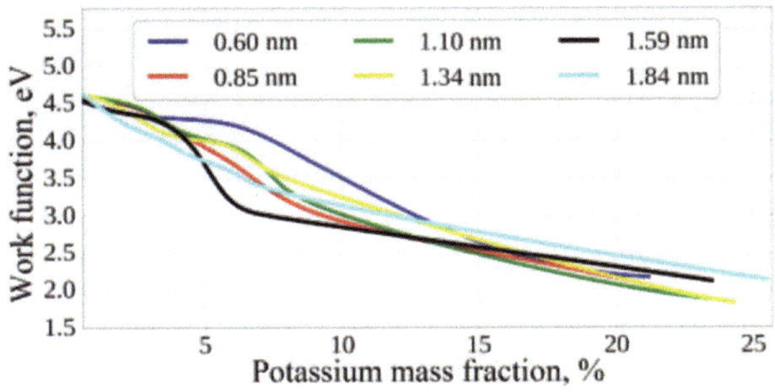

Figure 4.16 Work function values for different CNT lengths and potassium mass fractions in the system.

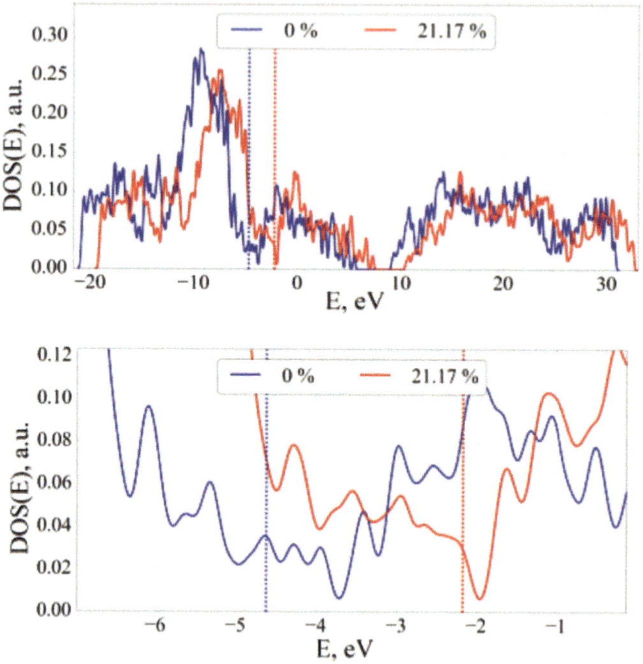

Figure 4.17 The density of electronic states for a pillared graphene monolayer film with a CNT length of 0.6 nm in the absence of potassium particles and with their maximum number of investigated.

between the work function of pure composite and the intercalated one can be ~2.4 eV. Figure 4.16 shows the graph of the dependence of the work function on the different mass fractions of potassium atoms in the system.

For all composites, the densities of electronic states were calculated. Figure 4.17 shows a graph of the density of electronic states for a monolayer composite with the shortest CNTs in its composition.

This graph shows that in the energy region near the Fermi level for a supercell with the shortest CNT, electron states that can be filled are allowed, both in the absence of potassium particles and at their largest mass fraction of considered. A similar picture is observed for monolayer pillared graphene with other CNTs (Fig. 4.18).

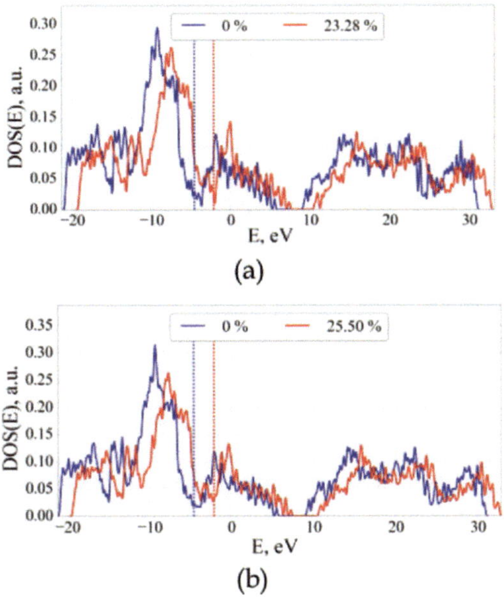

Figure 4.18 The density of electronic states for a pillared graphene monolayer with CNT lengths of 1.1 nm (a) and 1.84 nm (b) in the absence of potassium atoms and with their largest number (of the investigated). The dashed lines indicate positions of the Fermi level.

Table 4.2 shows the electron–energy characteristics of pure films of a monolayer pillared graphene and films of a pillared graphene with a maximum mass fraction of potassium atoms in the supercell.

The data from the table show that the emission properties of all films, depending on the potassium mass fraction, increase due to a decrease in the work function, but for some of them the number of available electronic states near the Fermi energy decreases, and it increases for others. With an increase in the CNT length, the density of states near the Fermi level for pure composites decreases; for potassium-saturated ones, it oscillates near the value 0.0297 without a pronounced decrease or growth.

Similar studies were carried out for bilayer pillared graphene films (Figs. 4.19 and 4.20).

The value of the transferred charge depends on the mass fraction of potassium in the supercell, and also on its location. In this

Table 4.2 Work function of a monolayer film of pillared graphene before and after the addition of potassium atoms

CNT length (nm)	0.60	0.85	1.10	1.34	1.59	1.84
Potassium atom mass fraction (%)	21.17	22.10	23.28	24.26	26.09	25.50
Charge transfer, electrons	12.973	13.309	15.723	16.764	17.122	17.234
Work function without potassium atoms (eV)	4.616	4.666	4.624	4.631	4.646	4.626
Work function (eV)	2.171	1.969	1.878	1.823	2.230	2.132
Work function decrease (%)	52.97	57.81	59.38	60.63	52.01	53.92
DOS (E_1)	0.0356	0.0337	0.029	0.0251	0.0305	0.0235
DOS (E_2)	0.0299	0.0262	0.0313	0.0299	0.0310	0.0298
DOS (E_2)/DOS (E_1)	0.840	0.777	1.079	1.191	1.016	1.264

Note: E_1, Fermi energy before intercalation; E_2, with the maximum mass fraction of potassium in the system.

regard, dependence in Fig. 4.21 can be explained as follows. In the presence of sufficiently large cavities in a bilayer film of a pillared graphene with long tubes, potassium atoms located near the carbon framework shield the rest of the potassium. Therefore, the charge distribution near the carbon framework is more active, far from it—significantly less. The value of the transferred charge from potassium atoms that are close to the composite can be about 0.95e; inside the cavity, it varies from −0.31 to 0.32 electron.

The work function is reduced; this value can be reduced by ~2.28–2.62 eV with a 12.46% mass fraction of potassium for a film with a CNT of 0.6 nm in length, 29.37% for a 1.10 nm CNT film, and 30.89% for a bilayer pillared graphene with a 1.84 nm CNT in its composition (Fig. 4.22).

Figure 4.23 shows the density of electronic states for a bilayer pillared graphene with different CNTs in its composition.

Graphs of the density of electronic states for pillared graphene bilayer films with different lengths of CNTs are similar; they are

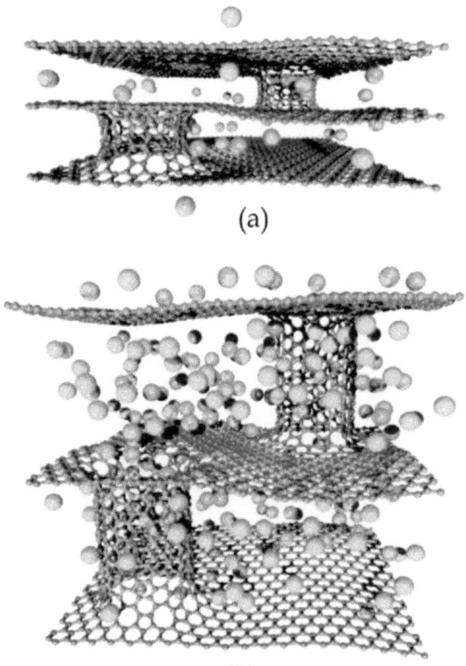

(a)

(b)

Figure 4.19 The supercells of the atomistic models of the pillared graphene bilayer, filled with potassium atoms: (a) 43 atoms; the length of the CNT is 0.6 nm; (b) 226 potassium atoms; the length of CNTs is 1.84 nm. Periodic boundaries are not shown.

topologically identical. However, there are differences in the density of electronic states for a two-layer columnar graphene near the Fermi level for different mass fractions of potassium; for all cases with a maximum potassium fraction in the system, the density of electronic states is higher.

Table 4.3 gives data about the supercells of the bilayer pillared graphene studied at the maximum mass fraction of potassium atoms in the supercell in comparison with the cells of pure films.

An increase in the density of electronic states allows us to state that the emission properties of intercalated bilayer pillared graphene are higher not only because of the decrease in the work function, but also because a larger number of electrons will be near

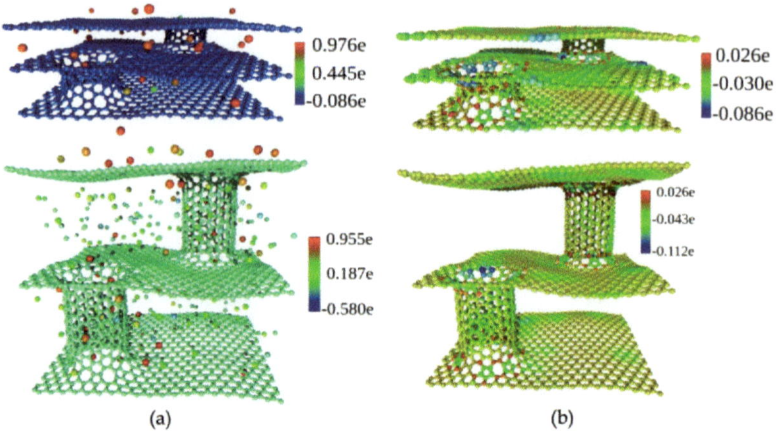

Figure 4.20 Charge distribution in supercells of the pillared graphene bilayer models filled with potassium atoms: (a) potassium atoms are present to form a gradient pattern; (b) potassium atoms do not contribute to the gradient picture. Periodic boundaries are not drawn.

Table 4.3 Work function of a bilayer film of pillared graphene before and after the addition of potassium atoms

CNT length (nm)	0.60	1.10	1.84
Potassium atom mass fraction (%)	12.47	29.37	30.89
Charge transfer, electrons	65.451	64.312	70.506
Work function without potassium atoms (eV)	4.505	4.507	4.506
Work function (eV)	2.223	1.887	2.099
Work function decrease (%)	50.65	58.13	53.42
DOS (E_1)	0.0132	0.0126	0.0125
DOS (E_2)	0.0646	0.1095	0.1102
DOS (E_2)/DOS (E_1)	4.894	8.690	8.816

Note: E_1, Fermi energy before intercalation; E_2, with the maximum mass fraction of potassium in the system.

the Fermi level. With an increase in the length of CNTs, an increase in the density of electronic states near the Fermi level is observed for composites with potassium.

For the 3D pillared graphene, the same values were calculated. The general trend, consisting in the charge transfer from potassium

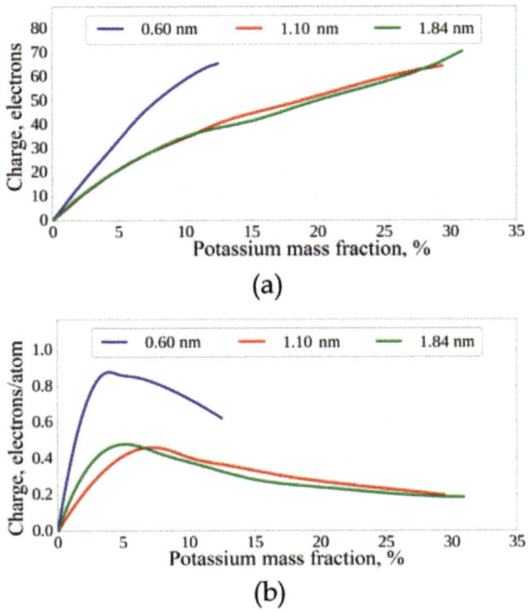

Figure 4.21 Dependence of the transferred charge from potassium atoms to the composite supercell of a pillared graphene bilayer at various mass fractions of potassium in the system: total (a), normalized to the number of potassium atoms (b).

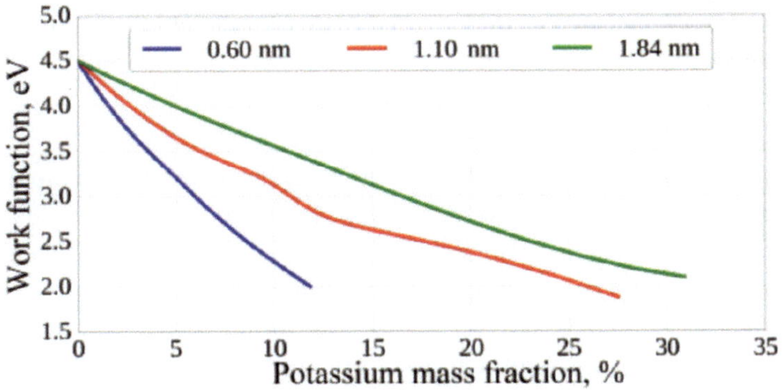

Figure 4.22 Dependence of the work function of a pillared graphene bilayer at various mass fractions of potassium in the system.

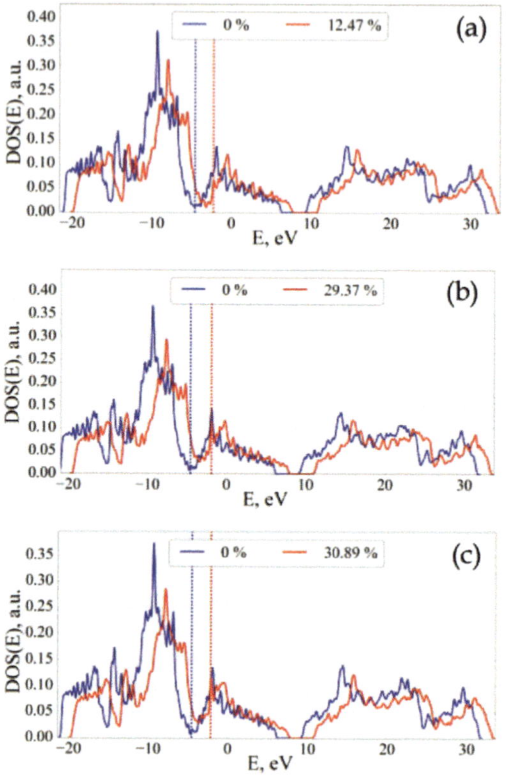

Figure 4.23 The density of electronic states for a pillared graphene bilayer with a CNT length of 0.6 nm (a), 1.10 nm (b), and 1.84 nm (c) for a different number of potassium atoms. The dashed lines indicate the positions of the Fermi level.

atoms to a carbon composite, accompanied by a decrease in the work function of the material, remains (Fig. 4.24).

The work function can be reduced by ~2.27–2.49 eV with a mass fraction of randomly arranged potassium atoms of 15.04% for a film with a CNT of length 0.6 nm, 25.46% for a film with a CNT of length 1.10 nm, and 36.43% for a 3D pillared graphene with a 1.84 nm tube in it (Fig. 4.25).

Figure 4.26 shows graphs of the density of electronic states for 3D pillared graphene depending on the length of CNTs in the composite and the mass fraction of potassium in it. Only closest to

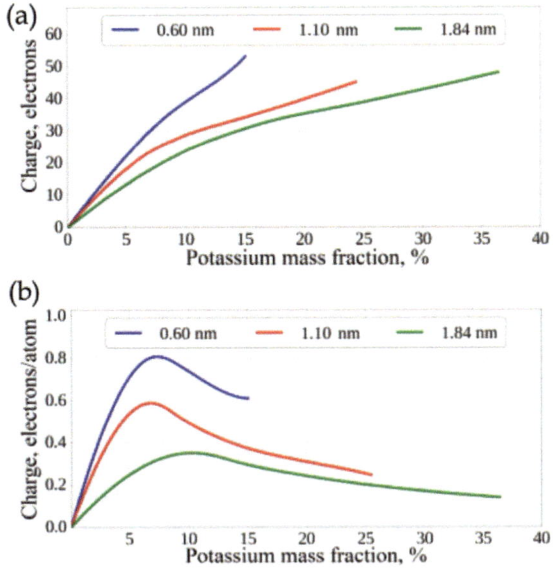

Figure 4.24 Transferred from potassium atoms to the cell of a 3D pillared graphene charge value at various mass fractions of potassium in the system: total (a), normalized to the number of potassium atoms (b).

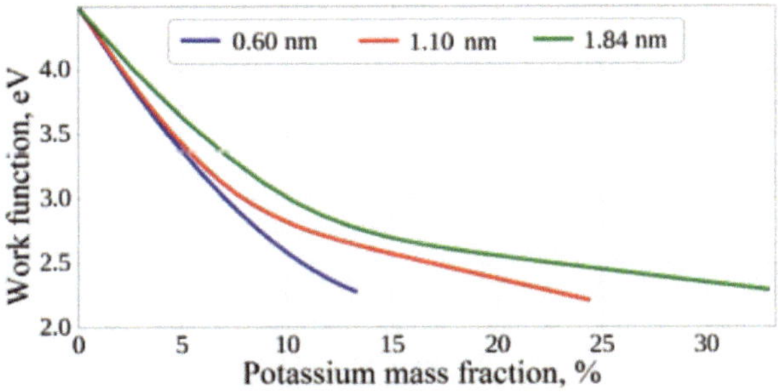

Figure 4.25 Dependence of the work function value from the number of potassium atoms in the system.

Fermi level area is shown, since the general appearance of all graphs is quite similar.

Table 4.4 gives the data about the supercells of 3D pillared graphene under consideration with the maximum mass fraction of

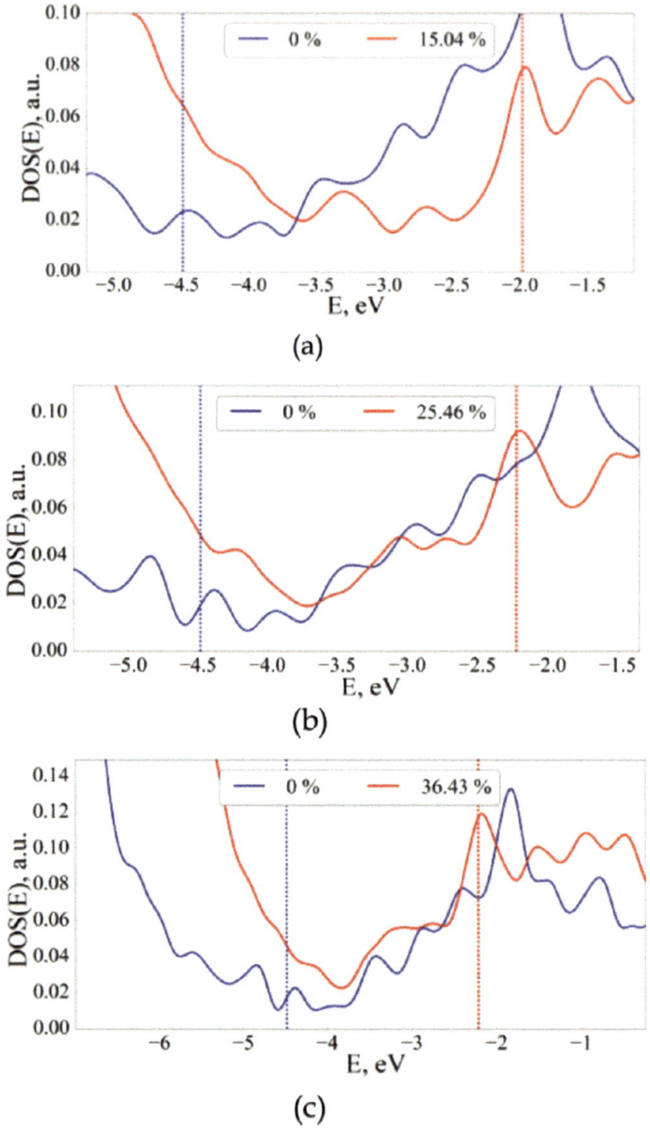

Figure 4.26 The density of electronic states for 3D pillared graphene with a CNT length of 0.6 nm (a), 1.10 nm (b), and 1.84 nm (c) for different mass fractions of potassium. The dashed lines indicate positions of the Fermi level.

Table 4.4 Work function of 3D pillared graphene before and after the addition of potassium atoms

CNT length (nm)	0.60	1.10	1.84
Potassium mass fraction (%)	15.04	25.46	36.43
Electron transfer, electrons	53.172	45.009	48.101
Work function without potassium atoms (eV)	4.505	4.507	4.506
Work function (eV)	1.983	2.222	2.209
Work function decrease (%)	55.98	50.70	50.98
DOS (E_1)	0.0231	0.0191	0.0173
DOS (E_2)	0.0783	0.0917	0.1173
DOS (E_2)/DOS (E_1)	3.390	4.801	6.780

Note: E_1, Fermi energy before intercalation; E_2, with the maximum mass fraction of potassium in the system.

potassium atoms in the supercell in comparison with the cells of pure structures.

4.4 Conclusion

The emission characteristics of monolayer, bilayer, and 3D pillared graphene are studied depending on the length of CNTs (which are part of the supercell), as well as the mass fraction of potassium atoms in the pores and cavities of the composite.

For pillared graphene cells without potassium atoms, the work function and density of electronic states were calculated depending on the cell type and the length of CNTs in it. It was found that the work function in bilayer films is below the analogous value of single-layer films by 0.129 eV on average. In a 3D composite, the work function is lower by $\sim$0.022 eV of the work function in bilayer films. The density of electronic states is higher in monolayer films, and the lowest in bilayer films.

For monolayer films with potassium atoms, the heat of formation was calculated; the obtained values showed stability of such composites.

For all the considered systems, a significant decrease in the work function of the material was observed with an increase in the number of added potassium atoms. At the same time, a significant decrease in the work function (⩾1 eV) is observed for all the systems considered with a ~10% mass fraction of potassium. A further increase in the number of added potassium atoms leads to an even greater decrease in the work function, more than 50% relative to the initial value (~2.5 eV).

The behavior of the density of electronic states is ambiguous in the case of monolayer pillared graphene films, in which a decrease in this value is observed near the Fermi level for composites with two smallest CNTs and an increase for the remaining structures in question (potassium-free cells). For bilayer composite and 3D composites, the density of electronic states on the Fermi energy increases with the addition of potassium atoms, from three times (3.390) to more than eight times (8.816). This indicates that the emission properties of materials are also increased because of the larger number of electrons near the Fermi energy.

References

1. Ni, Z., Ishaq, A., Yan, L., Gong, J., and Zhu, D. (2009). Enhanced electron field emission of carbon nanotubes by Si ion beam irradiation, *J. Phys. D Appl. Phys.*, **42**, pp. 075408-1–075408-4.
2. Kim, D. H., Lee, H. R., Lee, M. W., Lee, J. H., Song, Y. H., Jee, J. G., and Lee, S. Y. (2002). Effect of the in situ Cs treatment on field emission of a multi-walled carbon nanotube, *Chem. Phys. Lett.*, **355(53)**, pp. 53–58.
3. Chun, K. Y. and Cheol, J. L. (2008). Potassium doping in the double-walled carbon nanotubes at room temperature, *J. Phys. Chem. C*, **112**, pp. 4492–4497.
4. Kim, J.-P., Chang, H.-B., Kim, B.-J., and Park, J.-S. (2013). Enhancement of electron emission and long-term stability of tip-type carbon nanotube field emitters via lithium coating, *Thin Solid Films*, **528**, pp. 242–246.
5. Izrael'yants, K. R., Orlov, A. P., Ormont, A. B., and Chirkova, E. G. (2017). Effect of the cesium and potassium doping of multiwalled carbon nanotubes grown in an electrical arc on their emission characteristics, *Phys. Solid State*, **59(4)**, pp. 819–824.
6. Zhao, G., Zhang, Q., Zhang, H., Yang, G., Zhou, O., and Qin, L. C. (2006). Field emission of electrons from a Cs-doped single carbon nanotube of known chiral indices, *Appl. Phys. Lett.*, **89**, pp. 263113-1–263113-3.

7. Datta, S. (2009). *Quantum Transport: Atom to Transistor, Regular and Chaotic Dynamics* (Institute of Computer Science; Izhevsk, Russia), pp. 516–527.

8. Glukhova, O. E., Kirillova, I. V., Savin, A. N., Grebenyuk, K. A., Slepchenkov, M. M., Kolesnikova, A. S., Fadeev, A. A., and Shmygin, D. S. (2014). Methods for increasing the emission capacity of carbon nanotubes, *Izvestiya SSU. New Episode. Series "Physics,"* **14(2)**, pp. 18–22.

Chapter 5

Application of 3D Porous Graphene in Emission Electronics: New Atomistic Model and Numerical Estimations

Olga E. Glukhova[a,b] **and Michael M. Slepchenkov**[a]

[a]*Physical Department, Saratov State University, 83 Astrakhanskaya Street 410010, Saratov, Russia*
[b]*I. M. Sechenov First Moscow State Medical University, Laboratory of Biomedical Nanotechnology, Trubetskaya Street 8-2, Moscow, 119991, Russia*
glukhovaoe@info.sgu.ru

This chapter is devoted to the computerized study of the process of functionalization of a new structural modification of glass-like carbon by potassium and oxygen. Energy-efficient methods of filling glass-like carbon with potassium and oxygen are determined. New regularities of changes in the electronic and emission properties of glass-like carbon as a result of the functionalization are established.

5.1 Introduction

Carbon exists in many allotropic states. In addition to diamond and graphite, some well-known forms are nanotubes, graphene, and porous carbon, which is often regarded as glass like. Glassy carbon

2D and 3D Graphene Nanocomposites: Fundamentals, Design, and Devices
Edited by Olga E. Glukhova

ISBN 978-981-4800-41-9 (Hardcover), 978-0-429-20150-9 (eBook)
www.jennystanford.com

was found in 1957 in the form of fullerene-like densely packed mono- and few-layer "cups," chaotically located sp^2 nanostructures. For pores of sizes in the range of several nanometers, the density of material is rather small and equals $\sim$1.5 g/cm^3. In the end of the 20th century, glassy carbon began to be used for the production of matrix field emission cathodes providing a stable current [1]. The emitter of glassy carbon had a rounding radius of 2 nm. Modern electronics also uses glass-like carbon and glass-like carbon-based composites as a source of electrons. Subjected to ion bombardment by Ar^+ ions, glass carbon samples provide a low-temperature stable emission current [2]. A ceramic composite containing mesoporous nitrogen-doped carbon glass is a very promising material for the construction of cathode for solid-state lithium-ion batteries. The battery recharge time is reduced by six times [3]. Today, the multilevel arrangement of matrix cathode tips based on glassy carbon allows generation of rather large currents ($\sim 2 \times 10^9$ A/cm^2) [4].

However, the possibilities of glassy carbon in emission electronics are not exhausted, since at the moment the possibility of filling its nano- and micropores with a substance capable of lowering the work function has not been explored. On the other hand, the features of glassy carbon oxidation have not been studied to estimate its influence on the electronic structure. However, oxygen will most likely saturate the edges of the nanostructures as the active element forming the pores. In this situation, it is important to assess how much oxidation will reduce the emissivity of glassy carbon. Doping of glassy carbon with potassium has not been actively investigated, although attempts have been carried out to saturate the surface of glassy carbon with potassium in order to impart it hydrophilic properties [5].

In recent years, interest in porous carbon has increased after the development of electron microscopy, which discovered the features of the porous carbon structure and its unique mechanical properties [6, 7]. This allowed not only to build fairly accurate atomistic models of glassy carbon, but also to classify it according to topology. Earlier, glassy carbon was simply subdivided into graphitized (anisotropic material from multilayer graphene flakes, forming layers) and ungraphitized (randomly located graphite flakes) forms [8, 9]. Now graphitized glassy carbon is also divided into two classes [6]: Type I glass carbon mainly consists of randomly distributed curved layers of graphene fragments. Type II glass carbon contains self-organizing nanometer fullerene-like spheroids and 3D-disordered multilayered flake scales [6, 10]. In some recent works, porous glass-like carbon has been regarded as the bridge between few-layered graphene and fullerenes [11].

It should be also noted that in all processes of emission, the biggest role is played by the emitter's surface. Depending on the adsorbed impurities on it as well as depending on the topology of the surface, the emission properties deteriorate or improve. Moreover, impurities and topology determine the energy of the surface and the distribution of the electron charge density. As a result, the totality of these factors determines the work function and the electrical conductivity of the emitting surface.

In connection with the aforementioned factors, this paper studies in silico the regularities of porous glass-like carbon functionalized by oxygen, hydrogen, and potassium and its influence on the electronic and emission properties. Here, the regularities of the electronic structure are investigated along with the charge transfer in the nanopores of carbon films when the pores are modified with alkali metal atoms of potassium. As a carbon nanoporous film, we considered a film with a thickness of 4–5 nm, the surface of which is represented by a system of nanopores formed by fragments of mono- and few-layer graphene flakes and fullerenes.

5.2 Atomistic Model of 3D Structure and Film

5.2.1 3D Structure

As described in detail by Harris [10], the atomistic model was built by putting a certain number of fullerene fragments into a periodic box with a subsequent optimization process (optimization implies the reconfiguration of the system until the minimum of the total energy is reached under the given conditions). This is a time-consuming process since it requires multiple rebuilding of the fragmented system with a consistent increase in temperature

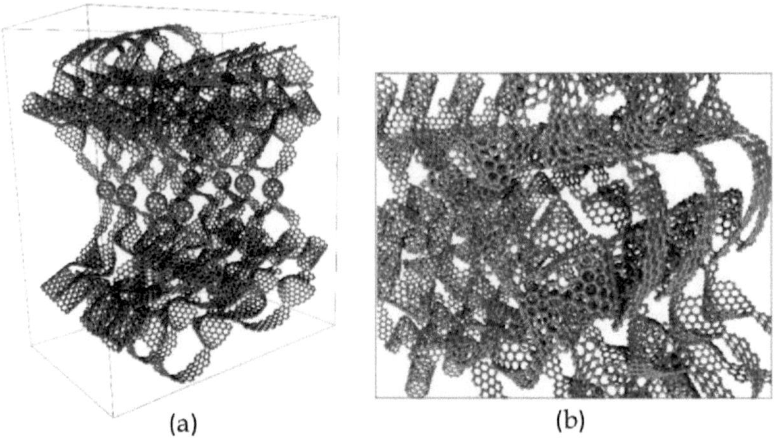

Figure 5.1 The atomic structure of the initial material for constructing the supercell of 3D structure: (a) view of the structure in the box; (b) atomic structure of the fragment.

and constant re-optimization. The process of construction was carried out by molecular dynamics in the AIREBO force field with the time step of 0.1 fs in the original software KVAZAR (http://nanokvazar.ru).

Initially we took fragments of mono-, bi-, and tri-layer fullerenes, nanotubes, graphene flakes and ribbons, nanocones, and nanotori fragments, which were randomly formed in the array that was placed in a box of dimensions 140 × 140 × 120 nm. This box contained 200 thousand atoms. The type of the box and the initial structure are shown in Fig. 5.1. Figure 5.1a shows a large box where all kinds of carbon nanostructures are located. It is clearly seen that this is only the initial structure, so it contains a large number of voids. A more detailed picture is shown in Fig. 5.1b. It is essentially the enlarged fragment of the structure placed in the large box. Further, the structure was optimized and subjected to uniform compression in all directions, which led to increase in the structure's density (i.e., the compression of the box deformed the structure).

Here, sequential heating of the structure up to 1300 K was conducted with the re-optimization of the atomic structure in the box. At the same time, the tendency to "fuse" the fragment edges was observed. To achieve this goal, carbon fibers and

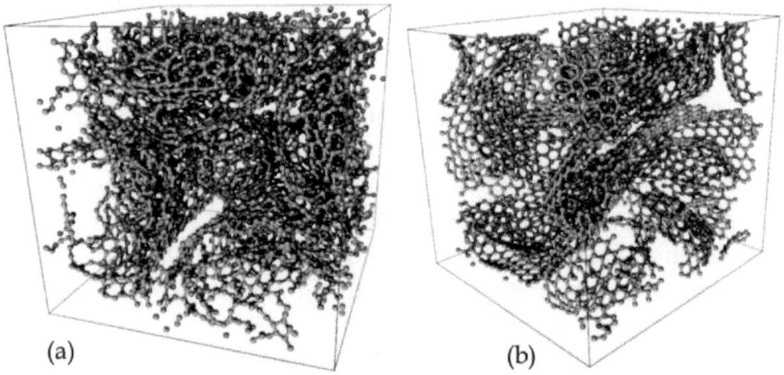

Figure 5.2 Atomic structure of the 3D supercell: (a) view of the structure in the box at the intermediate stage; (b) atomic structure of the 3D glass-like carbon supercell corresponding to a stable state.

other carbon structures were added to the area between the closest edges of fragments (single fibers of length 2–6 atoms, graphene flakes, isolated carbon hexagons). The appearance of defects such as pentagons and hexagons was monitored. As a result, a thermodynamically stable configuration was obtained in a cubic cell. This whole procedure was carried out until the density of the structure reached the density close to the density of glass-like nanostructures.

Optimization of the nanostructure in the large box was carried out for a long time. From the resulting structure, a cubic fragment with dimensions 5 × 5 × 5 nm was cut out and periodic boundary conditions were set. Despite the fact that glass-like carbon is not a periodic structure, we had to introduce a supercell to study the properties of this interesting material. Such a fragment in a small periodic box is shown in Fig. 5.2a. The structure represents an unordered mixture of carbon nanostructures whose atoms are in the state of sp-, sp^2-, and sp^3 hybridization. Since the atomic grid of the periodic box is chaotic, the translation of the box made the material isotropic in correspondence with the real situation. The view of the box structure corresponded with the structure after 15 ps of optimization at the temperature of 1300 K. After that, the structure was again optimized and insignificantly contracted in all directions.

During this process, the density of the structure increased. It was done to achieve the required density corresponding to similar porous carbons of type II glass carbon. In Ref. [6], its value equals to 1.42 g/cm^3. The density of our glassy carbon model is 1.22 g/cm^3 with the dimensions of the cubic cell 4 × 4 × 4 nm. The number of atoms in the supercell is 3918. The size of the box is typical for the modeling of porous glassy carbon since the pore sizes for type II glass carbon are 0.25–2.00 nm as shown by Harris [9, 10].

Further, a consecutive heating of the structure up to 1500 K was performed, followed by the optimization of the atomic mesh. At the same time, the tendency for the destruction and formation of new fragments was analyzed. Again, as before, carbon fibers and carbon structures were added to the region between the close edges of the fragments (atoms fibers, graphene flakes, isolated carbon penta-, hexa-, and heptagons). The appearance of defects such as pentagons, heptagons, and octagons was observed. As a result, a stable configuration in a cubic cell was obtained. The structure of the obtained model had three large cavities of 0.8–1.0 nm in size and 13 smaller pores with a size of about 0.2–0.5 nm. Then the amount of non-hexagonal elements was 10.6%, which also completely corresponded with the structure of type II glass carbon [10], which was characterized by 10–15% of non-hexagonal elements. To refine the atomistic model, re-optimization was performed by the SCC DFTB 2 method in the DFTB+ software (https://www.dftbplus.org). The cell corresponding to a stable system and a minimum of total energy is shown in Fig. 5.2b.

To understand the behavior of the obtained supercell under comprehensive compression, we investigated the dependence of the volume compression coefficient $K = -V \cdot dP/dV$ on the pressure. Figure 5.3 shows the periodic box with forces compressing the supercell along three axes and the dependence of the volume compression coefficient on pressure calculated by the AIREBO method.

Pressure was introduced only at the stage of building an atomistic model of a glassy carbon material. In our case, this was the pressure exerted on a cubic fragment of glassy carbon during the uniform compression of the object along all directions. Due to the pressure forces, we achieved the density of the considered object

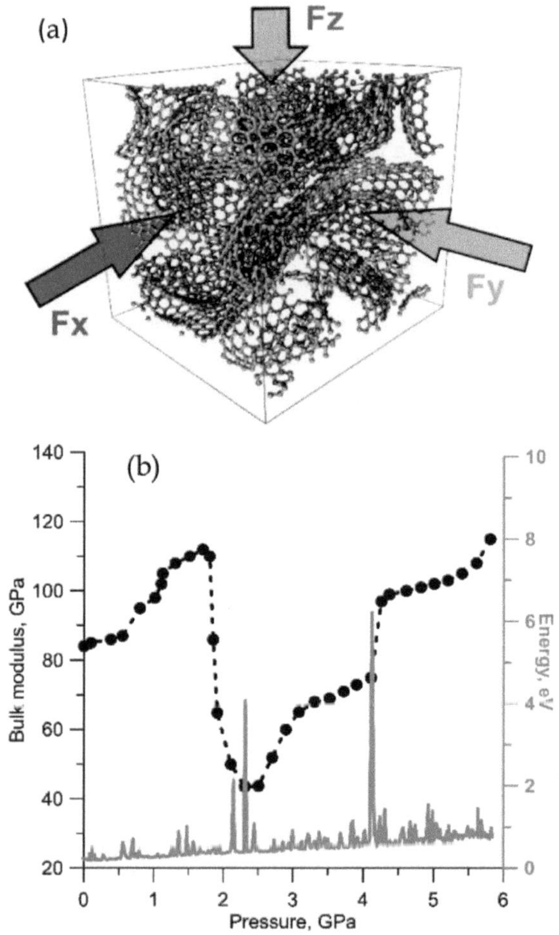

Figure 5.3 Comprehensive compression of the glassy carbon supercell: (a) structure of the fragment indicating compressive forces; (b) volume compression ratio and the energy of the structure as a function of the external pressure.

(~1.22 g/cm^3) close to the known data (~1.3–1.5 g/cm^3). Due to the pressure, the volume compression ratio K of the material was also calculated in order to compare the obtained value with the known results for glassy carbon of type II. The pressure exerted on a cubic cell of glassy carbon ranged from 0 to 6 GPa. First of all, we were interested in finding the phase transition point of the considered structure. From calculations, it was found that such a transition of the structure was observed at a pressure of about 2 GPa. From the presented graph (Fig. 5.3b), it can be seen that sharp changes in the coefficient K occur at a pressure near 2 and 4 GPa. These values of pressure correspond to sharp jumps in the structure's energy (the graph of the change in the total energy of the structure is marked by orange in Fig. 5.3b).

A complete coincidence between the values of the density and the coefficient K of the considered structure and the glassy carbon type II was not achieved, since our paper proposed a new structural modification of glassy carbon that does not directly relate to any of the two topological classes known at the present time: type I glass carbon and type II glass carbon. Our model is a kind of intermediate topological variant between these two classes. After the desired density of the glassy carbon structure was reached, the deformation effect was stopped. Further, a subsequent heating of the structure up to 1500 K followed by the re-optimization of the atomic mesh was performed. The influence of pressure on the mass density and the electronic structure was not considered.

For our model, the value K is several times higher than the values given in Ref. [6] for type II glass carbon. The difference can be explained by the difference of models in density and by the content of non-hexagonal elements and topology (in Ref. [6], the model is built only on multilayered graphene-like polyhedral structures containing fullerenes). It should be noted that the graphs of K versus pressure are in qualitative agreement with each other. Both for the sample in Ref. [6] and for our model, a noticeable decrease in the value of K is noted at the pressure of 2 GPa. In our case, further increase in pressure leads to a radical restructuring of the atomic structure, so it is not considered.

5.2.2 Film

The supercell of 2D porous carbon material was represented as a thin layer of glass-like nanomaterial, whose surface corresponded to the ground state. The technique of obtaining the supercell was the same. As a result, the density of the created film model was 1.21 g/cm^3 with dimensions of the supercell $4.2 \times 4.2 \times 4.1$ nm.

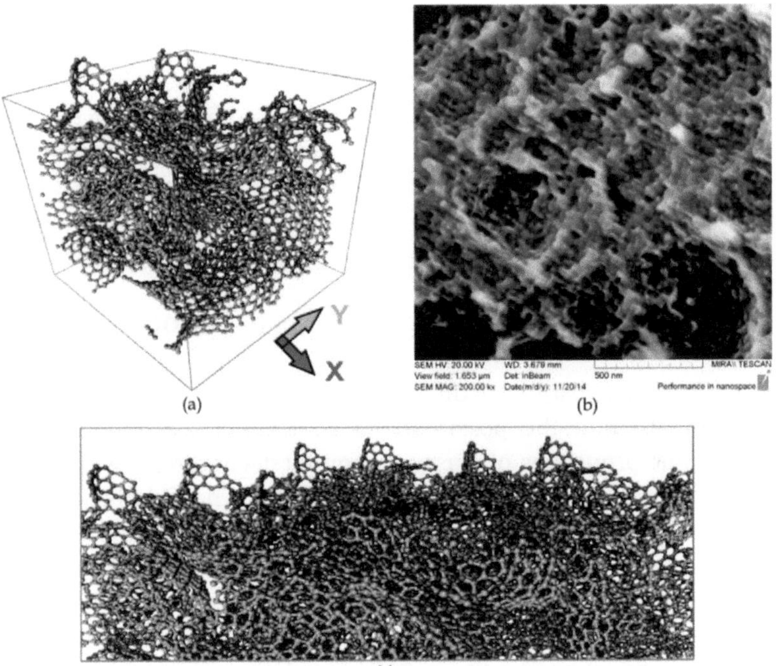

(a) (b) (c)

Figure 5.4 The surface of the porous glass-like carbon film is (a) the supercell of the film representing the surface of a glass-like nanomaterial; (b) image of glassy carbon surface fragment; (c) topology of the atomic model surface.

Since the film had a 2D structure, the supercell of the film was translated only in the x and y directions, as shown in Fig. 5.4a. The surface was a large number of protruding scales, making the surface rough and porous, which corresponded with the images of the surface (Fig. 5.4b). The number of atoms in the supercell was 3874. The size of the pores in our model was 0.25–1.00 nm, which completely corresponded with the glass-like porous nanomaterial. The number of hexagonal elements was 9.2%, which also completely corresponded with the structure of a glass-like porous nanomaterial. The supercell was translated in the x and y directions and reproduced to the surface of the synthesized glassy carbon in some approximation, as shown in Fig. 5.4c. The final atomic meshes were obtained by optimization with the AIREBO and the SCC DFTB 2 methods.

The developed surface of nanoporous carbon material model represented numerous fragments of graphene flakes and fullerenes. Nanopores were located on the surface, not deeper than 2.6 nm. Comparing the image in Fig. 5.4b with the atomistic surface of Fig. 5.4c, one can see the similarity that makes it possible to consider the built model of the film corresponding to the synthesized glass-like carbon.

5.3 Electronic Structure of Glassy Carbon

The functionalization of the glassy carbon was investigated with the aim to find the possibility of controlling the emission and electronic properties of glassy carbon by decorating it with atoms of various chemical elements. As it is well known, oxidation reduces the emissivity, while alkaline earth elements improve it. Therefore, we considered two cases of functionalization: by oxygen and by potassium. Potassium was chosen because the surface of glassy carbon was already successfully modified with potassium, as shown in the introduction. We also note that oxidation always occurs in the synthesis in one or another way, so it is necessary to know about its influence on electronic properties.

Initially, we calculated the band structure of pure glassy carbon. The band structure and energy were calculated by the SCC DFTB 2 method (http://www.dftb.org). Let us note that the choice of the SCC DFTB 2 method for calculating the electron structure of glassy carbon was made due to the large number of atoms (~4000 atoms) in the considered material cubic fragment. The speed of calculations is the main factor during considering such systems. As it is known, the SCC DFTB 2 method is 250 times faster than the DFT method with RI-PBE exchange-correlation functional and more than 1000 times faster than the DFT method with the B3LYP exchange-correlation functional. In contrast, the DFT method is usually used to calculate the electronic characteristics of low-atomic objects; therefore, it was not involved in our studies. To ensure high accuracy of the calculations of the band spectrums of pure glassy carbon and glassy carbon doped with potassium and oxygen, we selected the basic sets "pbc" and "3ob" provided by the official site of the

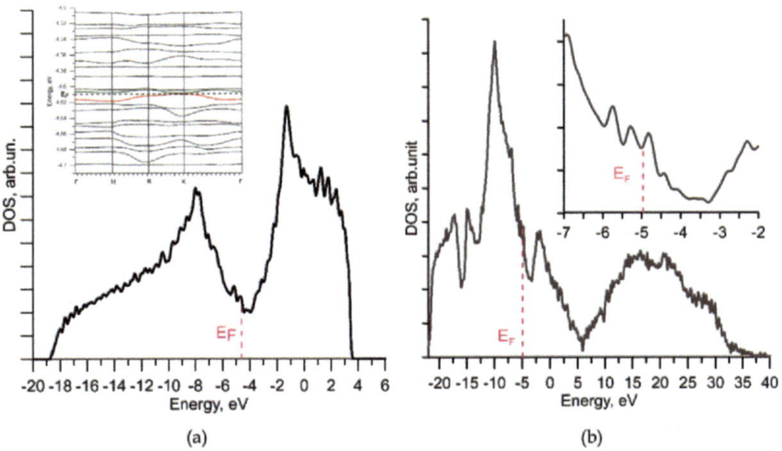

Figure 5.5 Energy structure of glass-like porous carbon: (a) band structure of pure 3D glass-like porous carbon (insert) and density of electronic states; (b) the density of electronic states of a glass-like porous carbon film (in the insert—DOS near the Fermi level). The dashed pick line shows the Fermi level.

DFTB+ development team (http://www.dftb.org/), implementing various mathematical models of DFTB. Earlier it was shown that the application of the SCC DFTB method allowed to predict correctly the electronic properties of various carbon nanomaterials, particularly graphene.

The distribution of the energy subbands near the Fermi level and the distribution of the density of electronic states (DOS) for 3D glass-like carbon are shown in Fig. 5.5. Red color marks the last subband of the valence band, and green color marks the first free subband of the conduction band. The Fermi level is $E_F = -4.61$ eV, which completely agrees with the experimental data on the work function of the glassy carbon. In Ref. [1], the value of the work function is indicated to be 4.6 eV. The high symmetry points of the Brillouin zone have coordinates $\Gamma(0, 0, 0)$, $M(\pi/a, 0, 0)$, $R(\pi/a, \pi/a, 0)$, $K(\pi/a, \pi/a, \pi/a)$. The value a corresponds to the length of the translation vector. Since the cube was taken as the unit cell, the translation vectors in all directions were the same in magnitude. The calculations showed that in the vicinity of K-point, the bands converged, indicating a high electronic conductivity of pure glassy carbon, which also matches the experimental data.

The DOS of the 2D glassy carbon film is shown in Fig. 5.5b. At the Fermi level, a high DOS is observed again. This suggests that both 3D and 2D glass-like porous structures are characterized by high electrical conductivity and serve as an excellent electrical conductor. This fact determines the high popularity of this material in emission electronics.

5.4 Oxidation of 3D/2D Glass-Like Carbon

The process of functionalization by oxygen was simulated by the addition of oxygen atoms to unsaturated bonds to the edges of fullerene-like structures and graphene scales. Then we optimized the structures by the SCC DFTB 2 method. The increase in the number of oxygen atoms was carried out uniformly throughout the periodic box with the condition that the oxygen atoms do not interact with each other. The interatomic distance between the oxygen and carbon atoms was 0.122 nm, which fit the known data about the C–O bond length. At each subsequent step, the heat of formation was monitored in order to ensure the exothermicity of the process. The dependence of the heat of formation on the amount of oxygen is shown in Fig. 5.6 (the required points are marked in red). To provide universality, the energy was recalculated per atom to represent models of 3D and 2D glass-like porous carbon on one graph. The maximum possible functionalization with oxygen was 7.05%. As expected, the oxidation of glassy carbon is not just an exothermic process, but also very favorable, since the energy appreciably decreases with increase in concentration.

At each step of the porous carbon functionalization by oxygen, the band structure was calculated by the SCC DFTB 2 method. The results of calculations showed that the oxidation of porous carbon in the considered limits did not change the conductivity properties. Conductivity remains close to that of metallic.

For example, for the 3D sample, the valence and conduction bands do not converge at the point *K* of the Brillouin zone, as it was in the case of pure glassy carbon in which they converge at the point

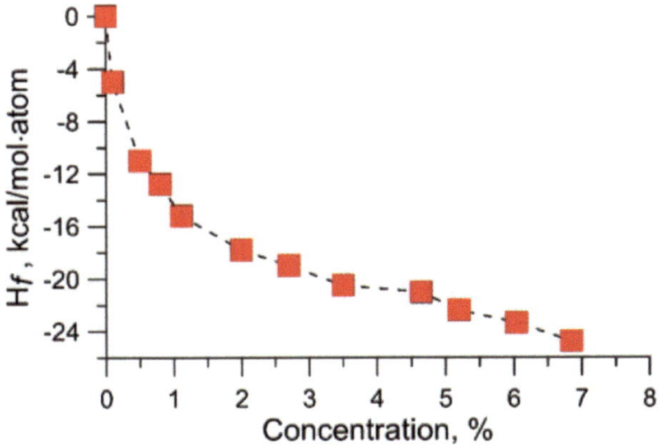

Figure 5.6 Heat of formation for functionalization of glass-like carbon by oxygen atoms.

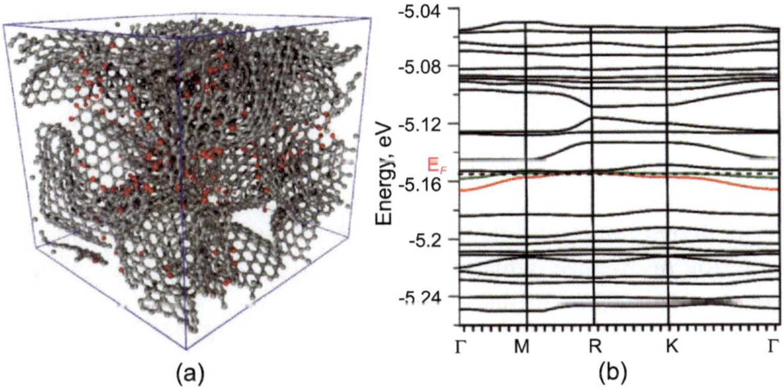

Figure 5.7 Elementary fragment of porous carbon: (a) atomistic model of porous glassy carbon with oxygen atoms (red points indicate oxygen atoms); (b) the band structure of oxidized 3D glassy carbon at the maximum concentration of 7.05% (the red line indicates the top of the valence band, and the green line indicates the bottom of the conduction band).

R. Figure 5.7a shows the cell of a porous glassy carbon functionalized by oxygen atoms in the maximum possible case for this atomistic model (7.05% oxygen concentration). Figure 5.7b shows the band structure of glassy carbon at the maximum oxygen concentration.

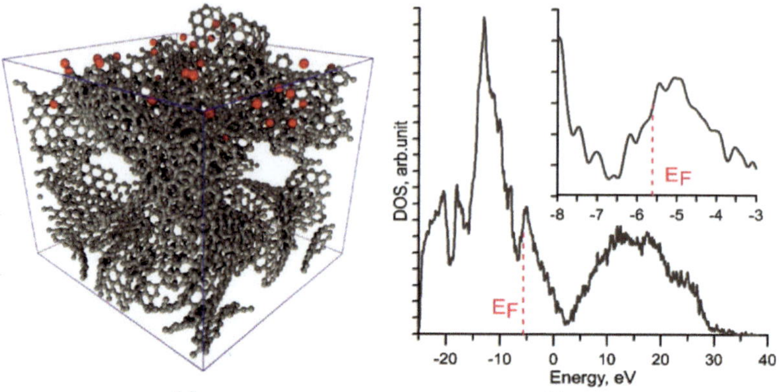

(a) (b)

Figure 5.8 Elementary fragment of porous carbon: (a) atomistic model of porous 2D glassy carbon with oxygen atoms (red dots mark oxygen atoms); (b) the DOS of oxidized 2D glassy carbon at the maximum concentration of 1.12%.

For a 2D sample, the patterns of oxidation looks different. Since oxidation occurs only on the sample's surface, the concentration of oxidation is small. The maximum concentration was only 1.12%. The distribution of oxygen atoms is shown in Fig. 5.8a. Figure 5.8b shows that the DOS for this concentration changed. However, the electrical conductivity did not decrease. The Fermi level changed its position shifting up to zero, which was expected since oxidation inevitably worsens emission properties.

5.5 Functionalization by Potassium

Functionalization by potassium was performed by filling the nanopores with potassium atoms. Carbon nanotubes were doped with potassium by precisely filling an inner cavity of a tube. In this case, the potassium atoms are located at the distance of ~0.31 nm from a wall of a carbon structure [12] and at the distance not closer than 4.6 nm from each other. We filled the pores with potassium in the same way as in the case of oxygen, gradually and evenly throughout the structure. The heat of formation was also monitored (Fig. 5.9). Filling the pores with potassium atoms was an exothermic

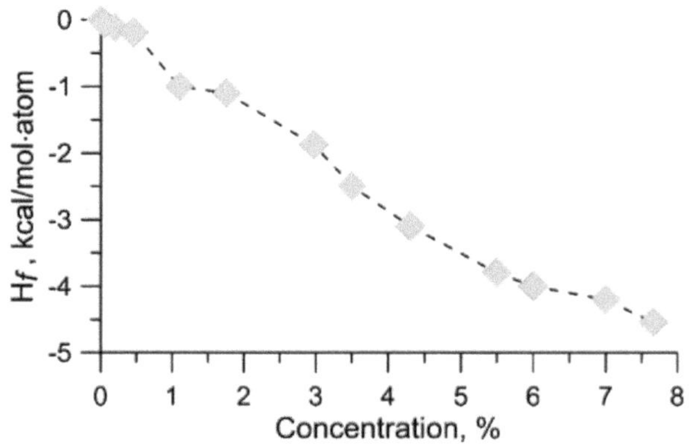

Figure 5.9 Heat of formation for functionalization of 3D and 2D glassy carbon with potassium atoms.

process. The maximum concentration of potassium in 3D glassy carbon equaled to 7.6%; in the case of glassy carbon film, it was only 4.65%.

Particular interest has been attached to the process of filling the nanopores on the film surface. The modification of the surface by potassium atoms was simulated by the molecular dynamics method MD SCC DFTB 2 under normal external conditions. The glassy carbon surface was bombarded with potassium atoms. The initial conditions were taken from Ref. [13] when the beam energy was set from 50 to 150 keV with a beam density of 10^{15}–10^{17} atom/cm^2. We considered the case when the beam density was 10^{16} atom/cm^2 and the energy was equal to 100 keV. Under such conditions, the atom velocity was 0.70 m/s. As the result of 20 numerical experiments, the main regularities of pores filling and reaching the maximum potassium concentration were found. It is established that for a given glass-like carbon surface, the maximum possible mass ratio of potassium was 4.65%.

During the bombarding of the glassy carbon surface, potassium atoms uniformly fill all the nanocavities. This is well demonstrated by the supercell of the 2D structure containing the maximum concentration of potassium (Fig. 5.11a). In this case, the potassium atoms uniformly fill either the surface or the deep and middle cavities (the potassium atoms are marked with blue-colored balls).

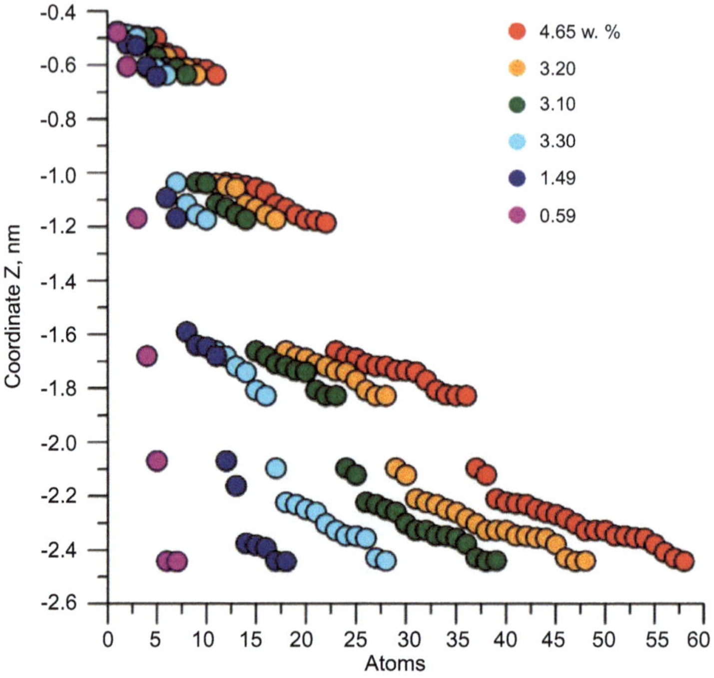

Figure 5.10 The distribution of potassium atoms in nanopores at different concentrations (mass ratio).

Thus, all pores at the depth of 2.5–2.6 nm are filled. The maximum degree of filling the 3D glassy carbon nanopores is shown in Fig. 5.11b.

Figure 5.10 shows the process of filling the nanopores with potassium atoms. The *y*-axis represents the depth of the nanopore (negative direction of the *z*-axis), and the *x*-axis represents the number of potassium atoms filling these nanopores. One can see the clear pattern of filling the glassy carbon surface nanocavities. Even at the low concentration of 0.59%, two potassium atoms reach the depth of 2.5 nm relative to the surface layer.

At each step of the porous carbon functionalization by oxygen and potassium, its band structures were calculated by the SCC DFTB2 method. We monitored the position of the Fermi level, which determined the work function of the material. The results

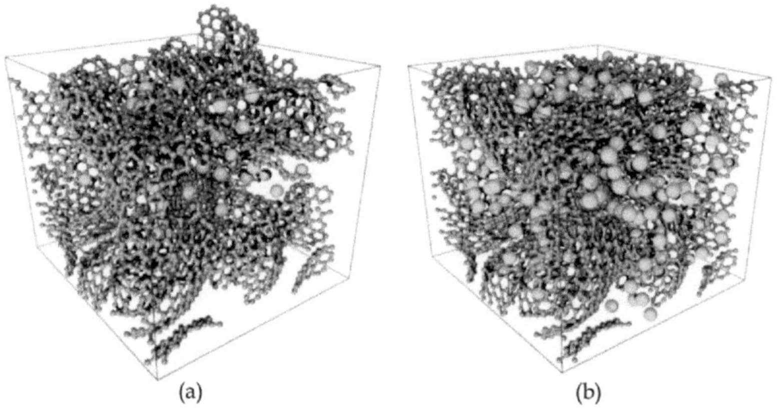

Figure 5.11 Filling the pores of glassy carbon by potassium: (a) in the case of glassy carbon film (concentration: 4.65%); (b) in the case of 3D structure (7.6%).

of calculations showed that the oxidation of porous carbon in the considered limits did not change conductivity properties. Figure 5.12a shows the calculated conduction band in the case of the bulk sample with maximum concentration of potassium (in the case shown in Fig. 5.11b). It is seen that during functionalization by potassium, the small gap appears in the region between the points R and K of the first Brillouin zone; this gap amounts only ~0.003 eV. However, the Fermi level in the case of maximum concentration is lowered by 1.27 eV. In practice, such effect can increase the field emission current by 10 times. Figure 5.12b shows the position of the Fermi level and the DOS of the glass-like carbon film. In this case, the Fermi level is not shifted appreciably and the surface modification slightly changes its position. This can be explained by the low concentration of potassium in the pores of the film surface in comparison to the bulk sample.

Let us continue the analysis of the DOS for various modifications of 2D glassy carbon. Figure 5.13a shows the DOS graphs for the oxidized surface with the oxygen concentration of 1.12% (red curve) and for the same surface at the maximum mass ratio of potassium (blue curve). It can be seen that at the Fermi level, the DOS of oxidized film with potassium is smaller in comparison to

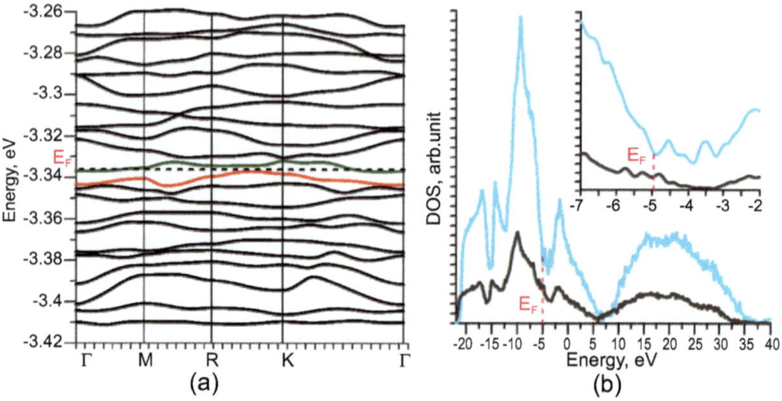

Figure 5.12 Electronic structure of porous carbon during functionalization by potassium: (a) zone structure of 3D glassy carbon; (b) the DOS of pure 2D glassy carbon (black line) and with maximum concentration of potassium (blue line).

the oxidized structure without potassium. The filling of oxidized film nanopores by potassium atoms is shown in Fig. 5.13b. For this case, it should be noted that the modification of surface pores with potassium does not give an appreciable result at the Fermi level, so we should not expect a noticeable improvement in the emission properties of this material. Another interesting case of pore filling is shown in Fig. 5.13c,d. Figure 5.13c shows the DOS for the case of a maximum upward shift in the Fermi level when potassium atoms fill pores of hydrogenated surface (the blue curve is the functionalization by potassium). It is clearly seen that in this case the DOS peak is located at the Fermi level, which improves electrical conductivity and reduces the work function. Green color corresponds to the DOS of the surfaces with 3.56% H, but without potassium atoms. The pattern of filling the hydrogenated surface by potassium is shown in Fig. 5.13d. The pores are filled equally as in the case of filling an oxidized surface.

We also calculated the DOS for two cases of 3D glass-like carbon functionalization, with the maximum concentrations of oxygen and potassium. The results are shown in Fig. 5.14. Figure 5.14a shows that the behavior of all three curves is the same. Therefore, the energy dispersion is very small. In order to understand how

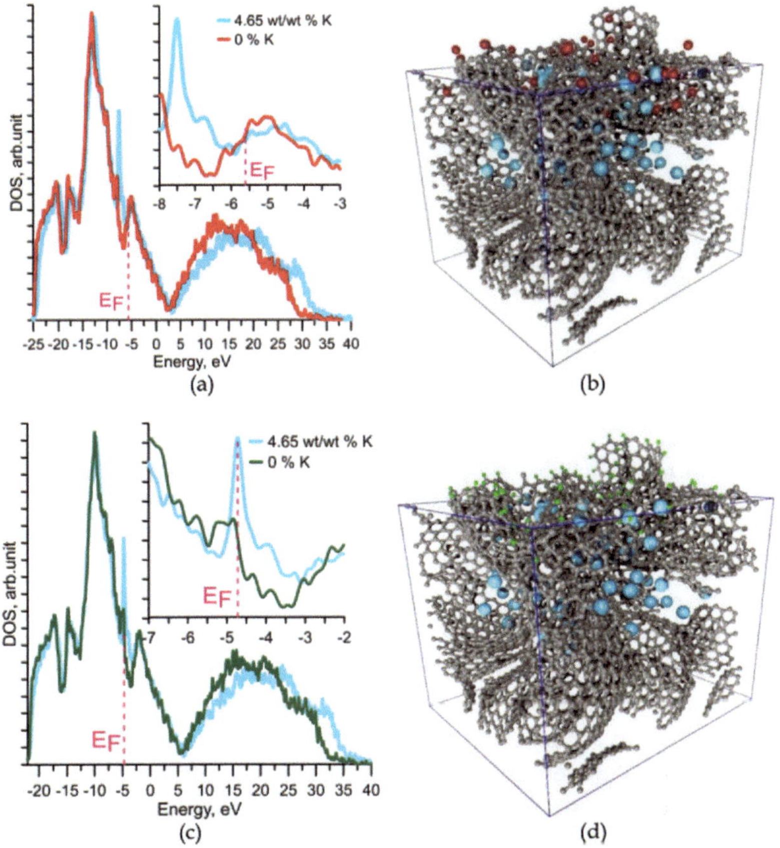

Figure 5.13 Modification of 2D glassy carbon: (a) oxidized surface (1.12%) with (blue curve) and without (red curve) potassium; (b) the supercell modified with potassium and oxygen atoms; (c) the DOS of hydrogenated surface (3.56%) with (blue curve) and without (red curve) potassium; (d) the supercell modified with hydrogen and potassium atoms.

the concentration of oxygen and potassium atoms affects the work function, we calculated the Fermi level at each stage of functionalization. This graph is shown in Fig. 5.14b. Blue balls mark potassium, and red squares mark oxygen. The given approximating curves show that in both cases, the Fermi level varies via quadratic law. It should be noted that among the modern carbon 3D nanomaterials that are actively used in the emission electronics, the

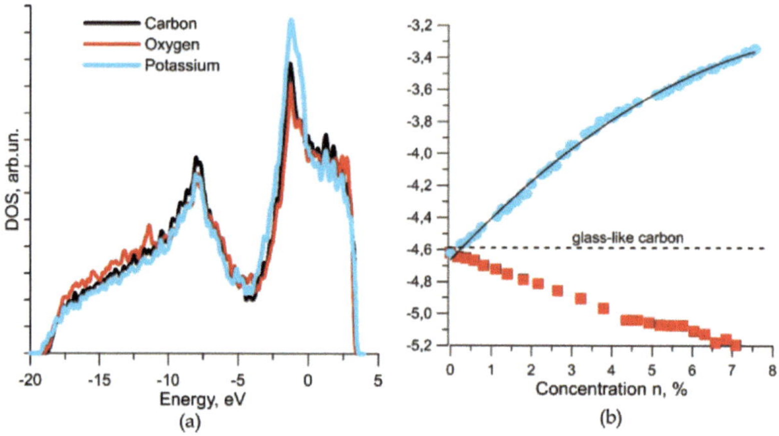

Figure 5.14 Functionalized glass-like carbon with the maximum concentrations of oxygen (red) and potassium (blue): (a) the DOS of pure glass-like carbon and functionalized glass-like carbon; (b) dependence of Fermi level on potassium and oxygen concentration.

glass-like carbon showed the highest sensitivity of the work function to the functionalization by alkali metal atoms. In particular, it was previously shown that the functionalization of carbon nanotubes by potassium atoms leads to a reduction in the work function up to 4.2–4.5 eV depending on the potassium concentration [14]. This is caused by some reasons, including the porous structure of the glass-like carbon material. Due to such structure, higher values of the maximum concentration of alkali metal atoms can be achieved in comparison with carbon nanotubes. Thus, on the basis of potassium functionalized glass-like carbon, it is possible to create a series of next-generation electronic devices that are characterized by high energy efficiency, stable modes of operation, small sizes, and long service life.

To find out the physical phenomena that occur during such changes in the film's energy structure and in the Fermi level, we have studied the charge transfer between potassium and carbon atoms with adsorbed atoms/atomic groups. Table 5.1 shows the charge values calculated by the Mulliken method. Charges localized on the carbon mesh, potassium, and other surface atoms are presented. The same cases of impurity atoms on the surface as in Fig. 5.13b,d are

Table 5.1 Distribution of charge for maximum concentration of potassium (4.65%)

	Charge, |e|				
N/N atoms, %	**K**	**H**	**O**	**OH**	**Carbon**
3.56% H	+39.95	+10.39	—	—	−50.34
2.22% H	+40.20	+7.10	—	—	−47.30
1.55% H	+40.46	+5.14	—	—	−45.60
0.81% O	+41.28	—	−10.38	—	−30.90
1.12% O	+41.45	—	−15.04	—	−26.41
2.58% OH	+40.10	—	—	−1.20	−38.90
1.12% O, 1.12% H	+40.75	+1.60	−8.42	—	−33.93

given. Atoms of potassium are always donors. Dependence of charge transfer from potassium atoms to the carbon mesh is nonlinear. An analysis of the DOS distribution showed that K atoms located near the framework lose almost a whole electron (0.8–0.9e), and K atoms located inside a group of potassium atoms practically lose nothing (the charge on these atoms is 0.26e). Therefore, the ratio of the charge to the number of potassium atoms in the cell varies from 0.86 to 0.73 e/atom.

Figure 5.15a shows the supercell with the maximum mass ratio of potassium (4.65%). The charges are displayed in color: the potassium atoms change color from red (charge +0.96e) to yellow-green (+0.56e) and green (+0.23e). The carbon mesh has a practically uniformly distributed charge −0.5e ÷ −0.23e; therefore, it is displayed in the green-blue color. Calculating the carbon mesh charge in all the cases of surface modification shows that there is a clear dependence between charge and work function. The more the charge transfer, the more the work function. Figure 5.15b shows the graphs of the change in the work function of the film with increasing charge on the carbon mesh for various cases of filling nanopores with potassium. Different colors indicate different cases of potassium concentration in nanopores. Each point corresponds to a certain value of the charge transferred from the potassium to the carbon mesh. Thus, important data were obtained about the regularities of the charge transfer from potassium atoms to the carbon mesh in glass-like film surface. Changing the mass ratio

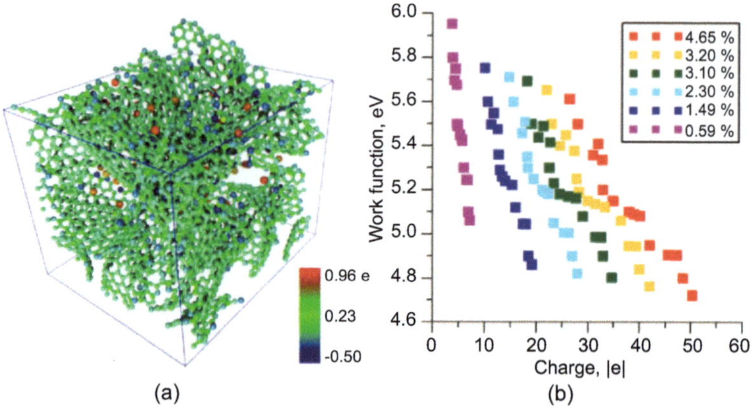

Figure 5.15 Charge transfer from potassium atoms to the carbon mesh of the film: (a) distribution of charge transfer on the surface of glassy carbon sample with 4.65% of potassium mass ratio; (b) dependence of work function on the charge transfer between potassium and carbon mesh for different concentrations of potassium.

of potassium atoms on the surface can reduce the work function even in the presence of impurity atoms that impair the emission properties. The analysis showed that the type of impurities in glass-like carbon films directly determined the nature of the change in energy structure and the position of Fermi level. Figure 5.16 shows the work functions on the potassium mass ratio for various variants of impurity atoms and its concentration. Typical groups of curves are distinguished. First, all cases with hydrogenation demonstrate decrease in the work function when nanopores are filled with potassium. As can be seen from Fig. 5.16, increase in the concentration of hydrogen atoms on the surface leads to a decrease in the work function with an increase in the potassium mass ratio. Second, regardless of the concentration of oxygen atoms, oxidation decreases the emission properties of the surface.

Regardless of the concentration of oxygen and potassium atoms, the work function always increases, in particular by 0.5–0.6 eV. Other variants of impurity atoms show a noticeable deterioration of the emission properties independent of concentration. Thus, the presence of OH groups, as calculations have shown, increases the work function by 0.5 eV in comparison to hydrogen. However,

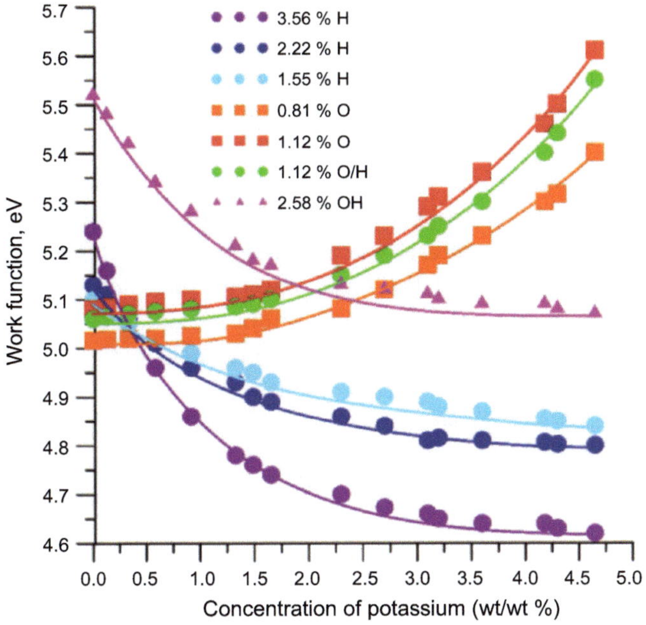

Figure 5.16 The change in the work function with an increase in the mass ratio of potassium at different concentrations of impurity atoms.

decoration with potassium can compensate it, as can be seen from Fig. 5.16 (curve with triangles). It should also be noted that the presence of both O and H atoms simultaneously on the surface as well as oxidation always contributes to increase in the work function.

5.6 Conclusion

From numerical simulations, we obtained new data about the electronic structure and emission properties of functionalized glass-like carbon in the form of 3D samples and 2D films with a developed surface. In the course of the study, we revealed the influence of oxygen, hydrogen, and potassium concentration on the electronic structure of glass-like carbon material, in particular, on a Fermi level that determines the work function. We found the limiting concentrations of potassium atoms filling (a) all nanopores of the bulk material and (b) the surface pores of the film. They are 7.66% for the bulk material and 4.65% for the film, respectively. The greatest decrease in the work function was observed in the bulk material at a potassium concentration of 7.66%, when the Fermi level increased by 1.27 eV and the work function decreased by 3.3 eV.

The considered types of functionalization are easily realized in practice, which predetermines a wide application of the proposed method for controlling the electronic and emission properties of glass-like carbon for the design of various nanodevices. In particular, there are wide prospects for developing field emission cathodes based on functionalized potassium glass-like carbon.

References

1. Sohda, Y., Tanenbaum, D. M., Turner, S. W., and Craighead, H. G. (1997). Fabrication of arrayed glassy carbon field emitters, *J. Vac. Sci. Technol. B*, **15(2)**, pp. 343–348.
2. Andrianova, N. N., Borisov, A. M., Mashkova, E. S., Nemov, A. S., Parilis, E. S., Sorokin, A. I., and Virgiliev, Yu. S. (2007). Ion-induced electron emission of glassy carbons, *Nucl. Instr. Meth. Phys. Res. B*, **256**, pp. 515–519.
3. Kichambare, P., Rodrigues, S., and Kumar, J. (2012). Mesoporous nitrogen-doped carbon-glass ceramic cathodes for solid-state lithium-oxygen batteries, *ACS Appl. Mater. Interfaces*, **4**, pp. 49–52.
4. Shesterkin, V. I. (2014). Effective emission area of multiple tip autoemission matrices made of glassy carbon, *J. Commun. Technol. Electron.*, **59(8)**, pp. 833–837.
5. Nakao, A., Iwaki, M., and Yokoyama, Y. (2003). Potassium ion implantation into glassy carbon, *Nucl. Instr. Meth. Phys. Res. B*, **206**, pp. 211–214.
6. Zhao, Z., Wang, E. F., Yan, H., Kono, Y., Wen, B., Bai, L., Shi, F., Zhang, J., Kenney-Benson, C., Park, C., Wang, Y., and Shen, G. (2015). Nanoarchitectured materials composed of fullerene-like spheroids and disordered grapheme layers with tunable mechanical properties, *Nat. Commun.*, **6**, 6212.
7. Bauer, J., Schroer, A., Schwaiger, R., and Kraft, O. (2016). Approaching theoretical strength in glassy carbon nanolattices, *Nat. Mater.*, **15**, pp. 438–443.

8. Franklin, R. E. (1951). Crystallite growth in graphitizing and nongraphitizing carbons, *Proc. Roy. Soc. Lond. A*, **209**, pp. 196–218.
9. Harris, P. J. F. (1997). Structure of non-graphitising carbons, *Int. Mater. Rev.*, **42(5)**, pp. 206–218.
10. Harris, P. J. F. (2013). Fullerene-like models for microporous carbon, *J. Mater. Sci.*, **48**, pp. 565–577.
11. Liang, H., Ma, X., Yang, Z., Wang, P., Zhang, X., Ren, Z., Xuea, M., and Chen, G. (2016). Emergence of superconductivity in doped glassy-carbon, *Carbon*, **99**, pp. 585–590.
12. Poklonskii, N. A., Kislyakov, E. F., Fedoruk, G. G., and Vyrko, S. A. (2000). Electronic structure model of a metal-filled carbon nanotube, *Phys. Solid State*, **42(10)**, pp. 1966–1971.
13. Nakao, A., Iwaki, M., and Yokoyama, Y. (2003). Potassium ion implantation into glassy carbon, *Nucl. Instr. Meth. Phys. Res. B*, **206**, pp. 211–214.
14. Blanchard, N. P., Hatton, R. A., and Silva, S. R. P. (2007). Tuning the work function of surface oxidised multi-wall carbon nanotubes via cation exchange, *Chem. Phys. Lett.*, **434**, pp. 92–95.

Chapter 6

Atomistic Structure and Electrical Conductivity of Pillared Graphene Films

Olga E. Glukhova[a,b] **and Dmitry Shmygin**[a]

[a]*Physical Department, Saratov State University, 83 Astrakhanskaya Street 410010, Saratov, Russia*
[b]*I. M. Sechenov First Moscow State Medical University, Laboratory of Biomedical Nanotechnology, Trubetskaya Street 8-2, Moscow, 119991, Russia*
glukhovaoe@info.sgu.ru

This chapter presents the results of a study of the atomic structure and electroconductive properties of films of a pillared graphene. A numerical estimate of the energy stability of the considered atomistic models of a pillared graphene with carbon nanotubes of various lengths and different diameters is given. The electron transmission function and electrical conductivity of mono- and bilayer films of a pillared graphene along the zigzag and armchair directions of graphene sheet are calculated.

6.1 Introduction

Pillared graphene is a carbon composite consisting of carbon nanotubes (CNTs) connected chemically to graphene sheets located perpendicular to the CNTs. A composite with orthogonal arrangement

2D and 3D Graphene Nanocomposites: Fundamentals, Design, and Devices
Edited by Olga E. Glukhova

ISBN 978-981-4800-41-9 (Hardcover), 978-0-429-20150-9 (eBook)
www.jennystanford.com

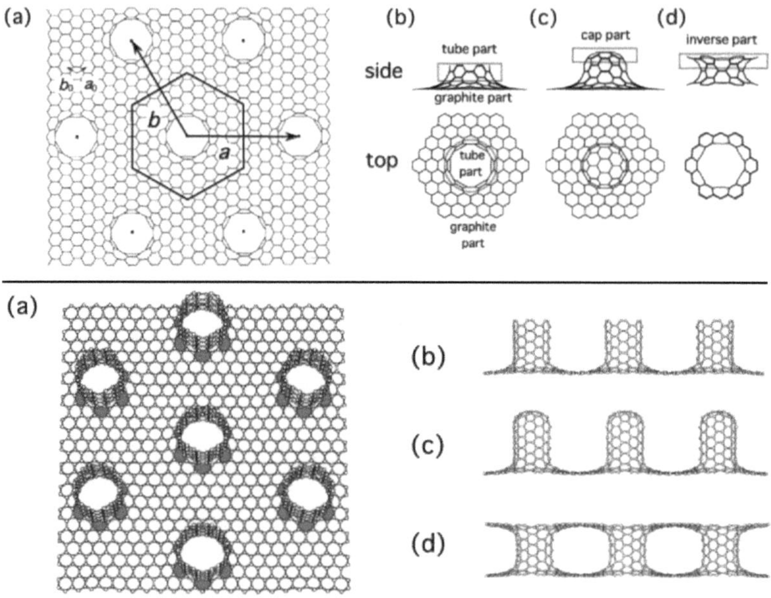

Figure 6.1 Pillared graphene models with a checkerboard arrangement of CNTs and one graphene sheet [5].

of graphene sheets is formed by the growth procedure for CNTs on suspended graphene. Initially, graphene is synthesized on a copper foil at the atmospheric pressure by chemical vapor deposition. Then, the graphene layers are transferred to a quartz substrate, and the formation of CNTs is observed at 750°C [1]. The pillared graphene production is mentioned in Refs. [2–4].

One of the first works on the properties of a similar material was released in 2002 [5]. The authors considered a similar scheme for the combination of CNT and graphene, but the material used was a single-layer quasi-2D structure in which CNTs of different lengths with open and closed ends were staggered on the surface of the graphene sheet (Fig. 6.1). Also structures were examined in which there were two graphene sheets. CNTs of the "armchair" type were considered, along with their chirality (6,6). The tight-binding method was used, and the band structure of such materials was calculated. However, calculations for structures completely analogous to those considered in this paper were not carried out; for the presented atomistic models, the electron transmission function (ETF) was not calculated, and the static electrical conductivity was not determined.

There are works devoted to the investigation of the heat-conducting [6–8] and mechanical [9] properties of pillared graphene of various configurations (with CNTs of different types and sizes), which showed high thermal conductivity and mechanical strength of the composite.

The strength of graphene–CNT junction was determined by the value of maximum stress that preceded the breaking of covalent bonds. This value was ~120 GPa for pillared graphene with CNT (6, 6). For connection with CNTs of the same diameter, but of a different chirality, a decrease in mechanical strength was observed. The calculated thermal conductivity in the plane of graphene sheets reached ~350–390 W/m/K, which indicates the high thermal conductive properties of composites that exceed the analogous characteristics for copper nanowires (70 W/m/K) [10].

Pillared graphene is actively used in the field of development of supercapacitors [11], since its resistance varies insignificantly for different bending radii.

Energy characteristics of a single-layer pillared graphene was also studied in Ref. [12], where various types of CNT and graphene compounds in a composite were considered. CNTs (6,6) were considered; it was shown that the presence of heptagons at the point of formation of bonds between CNT and graphene significantly influences the density of the electronic states of the material and it is the reason for the effect of size—the width of the band gap depends on the length of the CNTs in the composite.

In experimental studies, systems with dimensions of the order of 1 μm are often investigated. Only models of a much smaller size can be investigated via computer simulation at the atomistic level—dozens of nanometers. The dimensions are limited by the choice of potential and by what properties are investigated. However, the design of composites can be more complicated, since the main limitation is not the topological complexity of the model, but its size (Fig. 6.2).

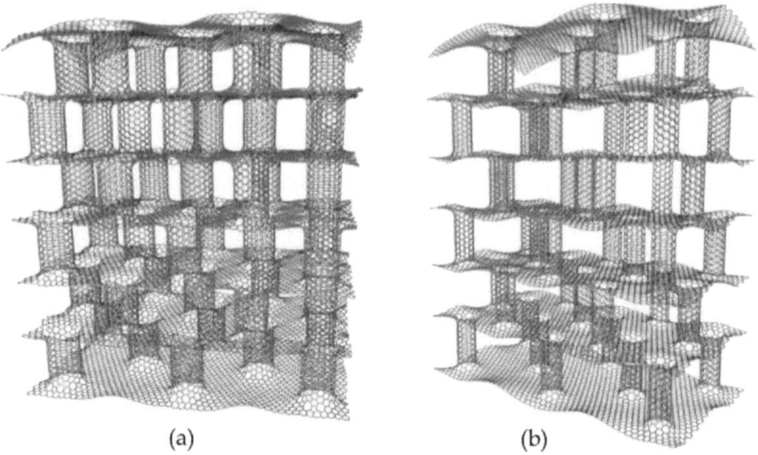

Figure 6.2 Atomistic models of pillared graphene, which include CNTs having diameters of 1.21 nm (a) and 0.8 nm (b), and a length of 2.09 nm. The numbers of carbon atoms in the models are 30510 and 28350. Sizes of models: length, 12.06 nm; width, 6.25 nm; height, 12.3 nm.

6.2 Atomistic Structure of Pillared Graphene

To investigate the properties of pillared graphene by a numerical experiment, we have studied the energy stability of models of different types. For further calculations, it means that an equilibrium (energetically stable) configuration of the initial structure participates in the simulation process. Stability was determined according to the value of the heat of formation—the difference between the sum of the energies of the constituent parts of the composite and the energy of the final structure:

$$\Delta U = U_{\text{opt}} - \sum_{i} U_i \tag{6.1}$$

In the case of formation of a stable configuration, this value is negative. Composites with different geometric parameters were considered: They differed in the diameters and lengths of the CNTs in them. At this stage, composites with two graphene sheets and one CNT (Fig. 6.3) were studied.

The process of studying the stability of composites can be divided into the following stages:

Figure 6.3 An example of a composite model consisting of two graphene sheets and one carbon nanotube with a length of 1.845 nm and a diameter of 1.212 nm.

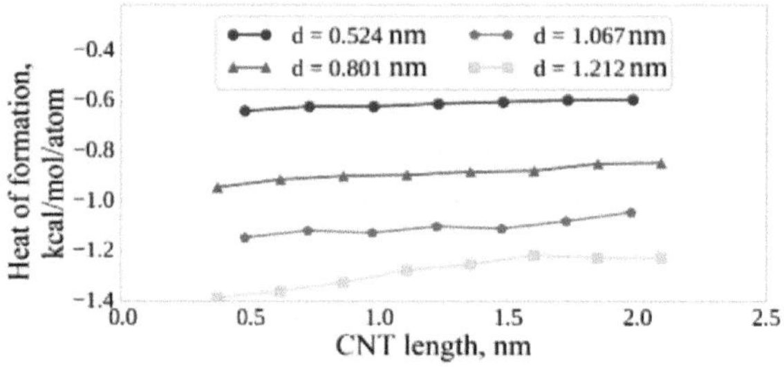

Figure 6.4 Dependence of the heat of formation on the length and diameter of CNTs in the composite.

1. Optimization (finding the equilibrium configuration of composite atoms, that is characterized by a minimum of total energy) of individual components of the system using the molecular dynamics with the Brenner potential [13, 14] and finding the values of the total energy for optimal configurations.
2. Calculation of the total energy of all parts of the system.
3. Optimizing the entire system and finding its total energy.
4. The heat of formation was determined as the difference between the total energy of the system and the sum of the energies of its components (Eq. 6.1).

For composites with tubes of different diameters (5.24 Å, 8.01 Å, 10.76 Å, 12.12 Å), the heat of formation was calculated as a function of the length of the CNT. The dependence of the heat of formation on the length of CNTs for different diameters of nanotubes is shown in Fig. 6.4 [15].

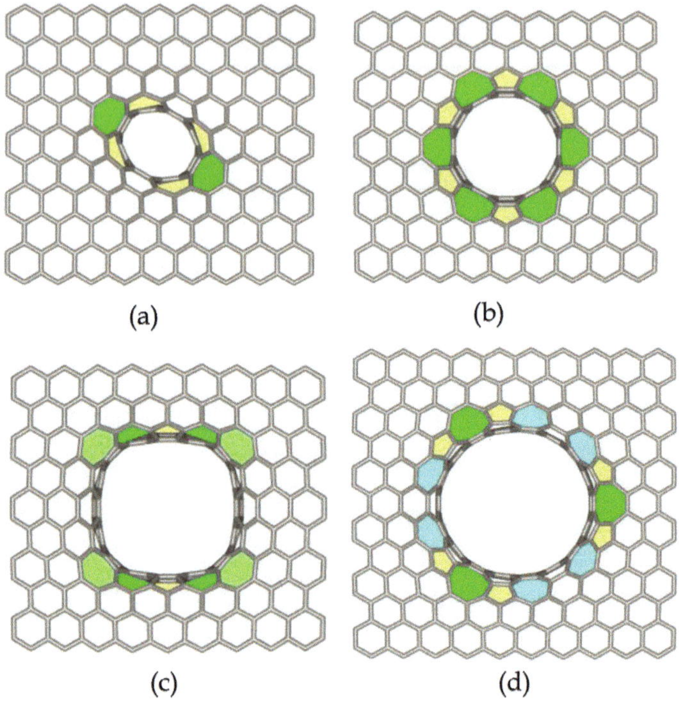

Figure 6.5 The junctions of CNT and graphene in pillared graphene with different CNT diameters: (a) 0.524 nm, (b) 0.801 nm, (c) 1.076 nm, (d) 1.212 nm. Color-highlighted defects in the form of non-hexagonal elements, which have 5, 7, and 8 carbon atoms.

After the optimization procedure, it was noted that the junctions of CNTs and graphene sheets contain defects located axially symmetrically (Fig. 6.5). The symmetry of the arrangement of the defects was preserved for all diameters. This fact was also noted in Ref. [16].

The data obtained make it possible to state that all presented composites represent a stable method of joining CNTs and graphene, with the energy stability increasing with an increase in the diameter of the CNTs in the composite. All the models considered were studied for energy stability without the use of periodic boundary conditions.

An additional study was conducted on the energy stability of pillared graphene models containing an array of CNTs and graphene

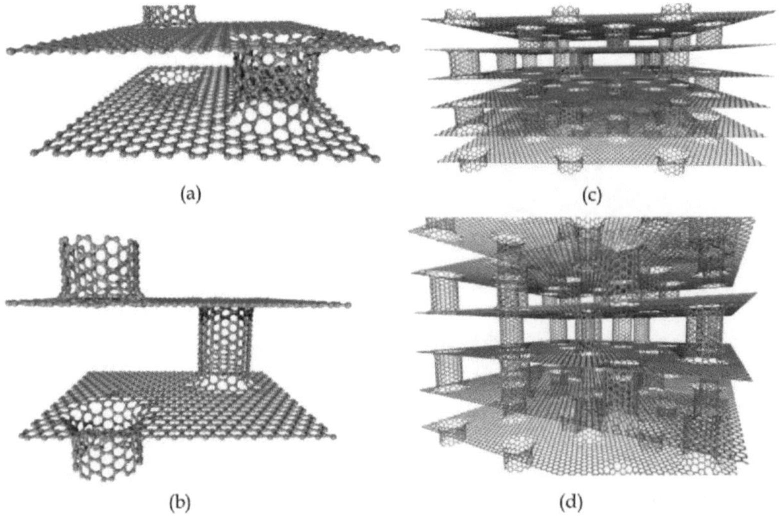

Figure 6.6 Supercells of 3D composites whose nanotubes have lengths of 1.102 nm and 1.844 nm (a, b), as well as examples of structures obtained after translation of the corresponding supercells (c, d) in three directions (integer coefficients in translation varied from 0 to 2).

sheets within the REBO potential. In particular, the composite, shown in Fig. 6.2a, is characterized by a heat of formation of −10.374 kcal/mol/atom, which indicates the stability of such a complex configuration. The most stable configuration is characterized by the maximum diameter of the examined CNTs. So in further calculations, the tube length was varied and the diameter was kept constant (~1.21–1.23 nm). The curvature of the graphene sheet is related to the period of the tube arrangement and also to the fact that the optimized model was considered with free boundary conditions and was not part of an infinite structure.

Figure 6.6 shows the models of supercells of 3D composites.

6.3 Atomic Structure of Pillared Graphene Films

In addition to the 3D composite, the properties of monolayer and bilayer films of pillared graphene were investigated. All investigated composites contain CNTs of the largest diameter from

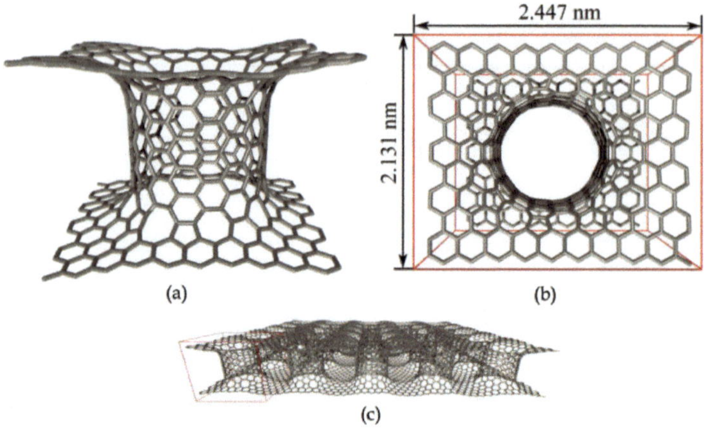

Figure 6.7 A monolayer film of pillared graphene (c), which contains carbon nanotubes with 1.23 nm of diameter and 1.102 nm of length, compiled on the basis of the supercell (highlighted in red in b, c), depicted in (a, b).

the considered—the "armchair" (9,9), because, according to the stability study conducted, they are the most stable. The distance between CNTs in all models was the same: 2.1 nm. Since a single-layer film can be composed of a sufficiently small cell (Fig. 6.7), a more accurate semi-empirical method can be used to model such structures, for example, DFTB [17].

The supercell of the composite shown in Fig. 6.8a includes 472 atoms. In Fig. 6.6, the smallest and the largest supercells of pillared graphene films consisting of 400 and 580 carbon atoms are shown.

Bilayer composites with three different lengths of CNTs were studied: 0.607 nm, 1.102 nm, and 1.844 nm. The numbers of atoms in the supercells of such films are quite large and are equal to 2400, 2544, and 2760 carbon atoms. Figure 6.9 shows cells of bilayer composites.

Despite the use of the calculation acceleration technique, the problem of finding the electrical conductivity of such structures is computationally complex and the calculation time depends nonlinearly on the number of supercell atoms. For this reason, the calculation of models containing longer nanotubes or having three or more layers remains computationally hard.

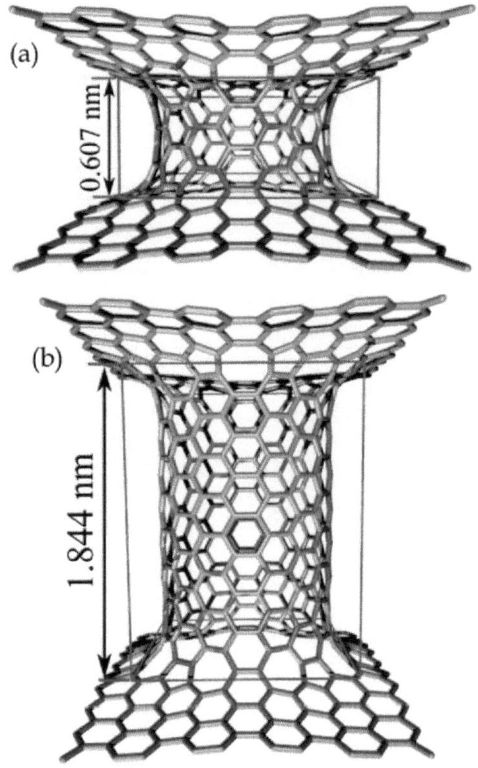

Figure 6.8 Supercells of pillared graphene films.

6.4 Transmission Function of Pillared Graphene Films

The ETF for monolayer pillared graphene films was investigated; the length of the tubes varied from 0.6 nm to 1.84 nm inclusive. The study was conducted along both translation directions: *x* and *y* (Fig. 6.10).

The average transmission function was obtained with the developed post-processing technique to reduce the calculation time.

In Fig. 6.11, the results for composites with all the examined CNT lengths are shown. The distance between the tubes was the same in all cases: 2.1 nm.

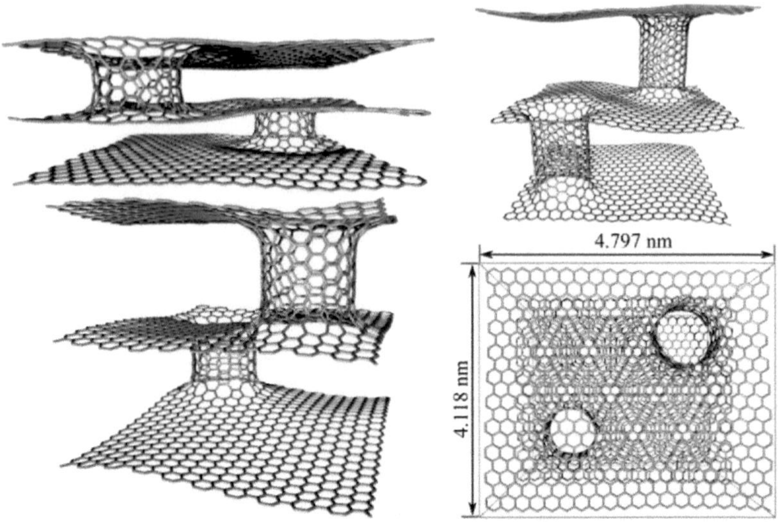

Figure 6.9 Supercells of bilayer pillared graphene films. The sizes along the x, y axes are the same for all cells and are indicated taking into account the periodic boundary conditions.

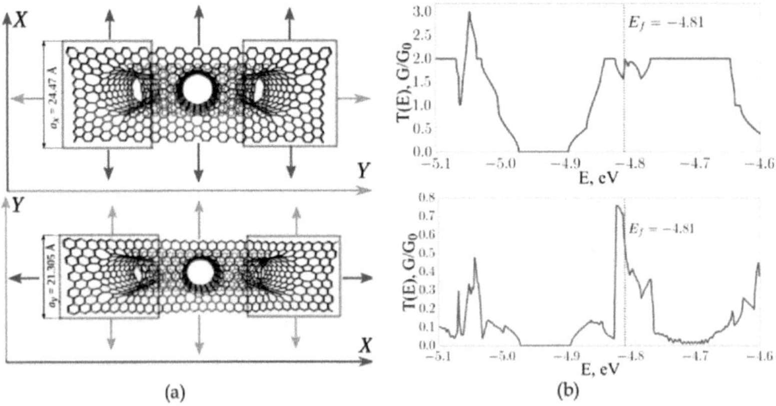

Figure 6.10 A film of pillared graphene with a carbon nanotube (9.9) (1.84 nm of length): (a) a supercell of the composite enclosed between electrodes, which are represented by similar cells, with electronic transport along the y direction (left) and along x (right); (b) the electron transmission function along the y direction (on the left) and along the x (on the right).

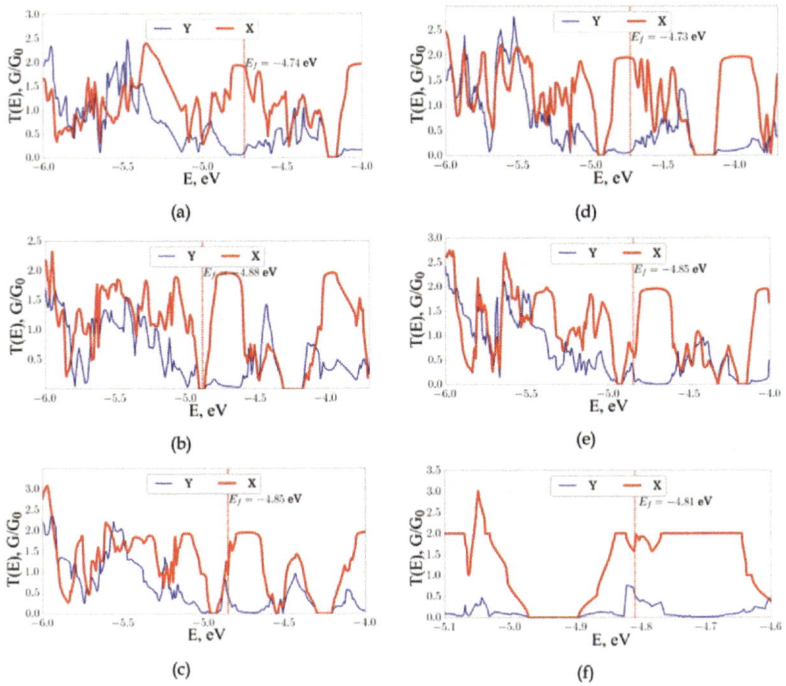

Figure 6.11 The electron transmission function for a monolayer film of pillared graphene at different lengths of carbon nanotubes in its composition: (a) $L = 0.607$ nm, (b) $L = 0.858$ nm, (c) $L = 1.102$ nm, (d) $L = 1.349$ nm, (e) $L = 1.598$ nm, (f) $L = 1.844$ nm. x, y is the direction of translation of the cell of the conducting channel.

It can be seen from the figure that along the edge of the "armchair" type, the electrical conductivity is higher in all the situations considered. The value of the Fermi energy for composites ranges from −4.88 to −4.73 eV. However, for each situation, the Fermi energy was calculated individually, because this value has a significant effect on the conductivity and resistance, which are calculated on the basis of the transmission function.

To understand the relief of the transmission function, additional calculations were carried out.

1. Graphene nanoribbons (GNRs) of "armchair" and "zigzag" type were considered. Sizes of the systems were chosen according to

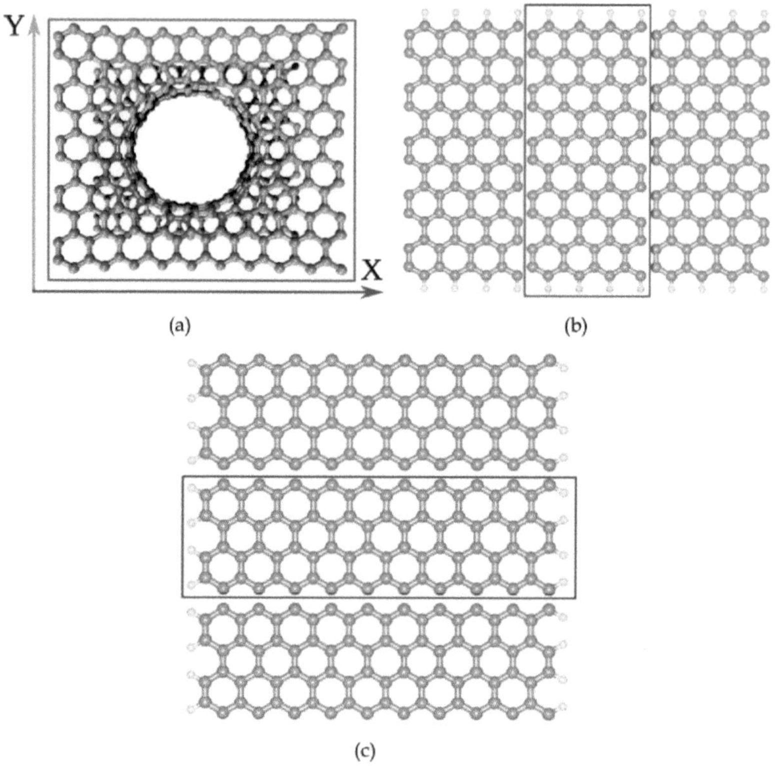

Figure 6.12 The supercell of pillared graphene (a), zigzag (b), and armchair (c) graphene nanoribbons, which are proportionate to a supercell in size.

the sizes of the supercell of the pillared graphene monolayer film (Fig. 6.12).

GNRs are 1D structures; their translation symmetry can be defined by one translation vector. The direction of current transfer in Figs. 6.12b,c for nanoribbons: parallel to x-axis (b) and y-axis (c). The ETFs for these models were calculated (Fig. 6.13).

From Fig. 6.13, it can be seen that graphene nanoribbons of the "zigzag" type are characterized by a higher value of the ETF at the Fermi level than the "armchair" nanotubes [18]. From the obtained values of the ETF, it is possible to obtain the values of the static electrical conductivity for the considered systems. They

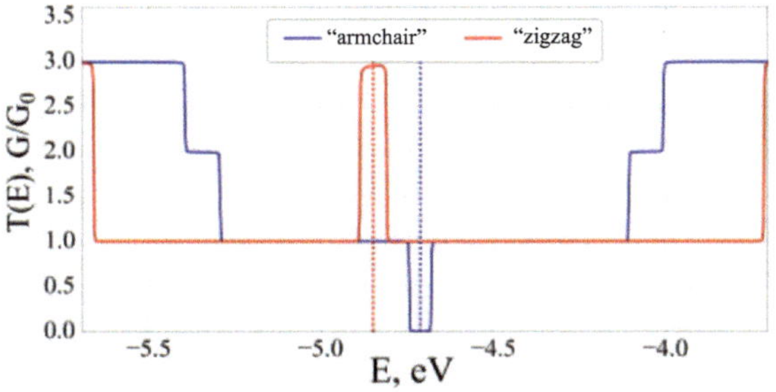

Figure 6.13 The transmission function of graphene nanoribbons of type "armchair" and "zigzag."

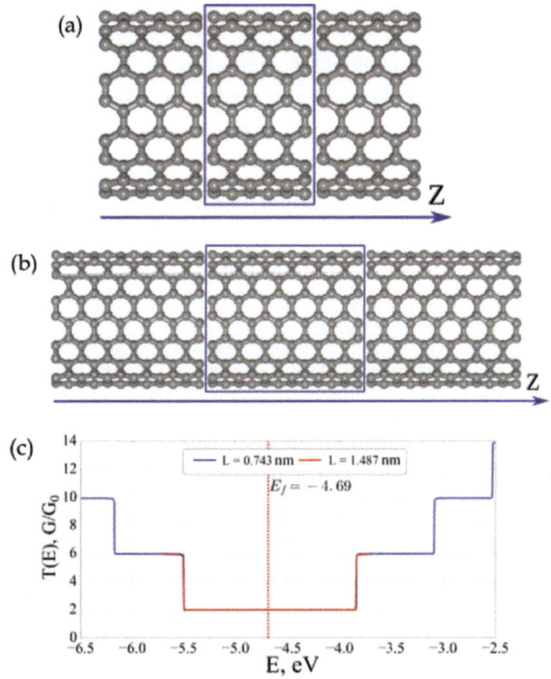

Figure 6.14 Atomistic models of CNTs (9,9) with lengths (a) 0.743 nm and (b) 1.487 nm. (c) Electron transmission function for the carbon nanotubes from (a, b).

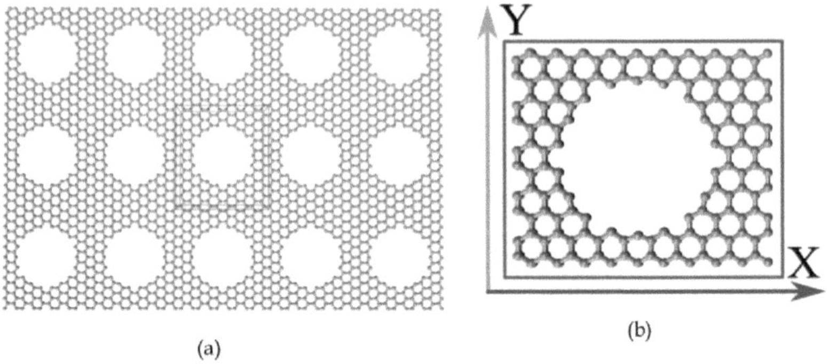

Figure 6.15 Atomistic model of porous graphene (a) and its supercell (b).

are equal to 35.081 μS for a nanoribbon of "armchair" type and 173.877 μS for a "zigzag"-type GNR.

2. The ETF was calculated for an infinite CTN of the "armchair" type with chirality (9,9). Since there is no dependence of the ETF values for the infinite CNT on the size of the cell, for two supercells of different lengths, the same values of the ETF were obtained (Fig. 6.14).

   Nanotubes of the "armchair" type shown in Fig. 6.14 have a metallic conductivity with two conduction channels at the Fermi level.

3. The ETF for a segment of a graphene sheet with an aperture, which was the unit cell of a porous graphene film (Fig. 6.15), was studied. The pores were located regularly.

   The pore size was determined by the diameter 1.23 nm of the CNTs (9,9), which were then used as a component of pillared graphene. An equilibrium configuration of the atomic structure was found using the conjugate gradient method, and the interatomic interaction was calculated according to the SCC DFTB method with the pbc-0-3 parametrization.

   The plots of ETF and the images of the atomistic models of porous graphene, which was used in simulations, are shown in Fig. 6.16.

   In Fig. 6.16, the influence of regular pores on the electrical conductivity of graphene films is noticeable. The most significant changes occurred with electronic transport along the "zigzag"

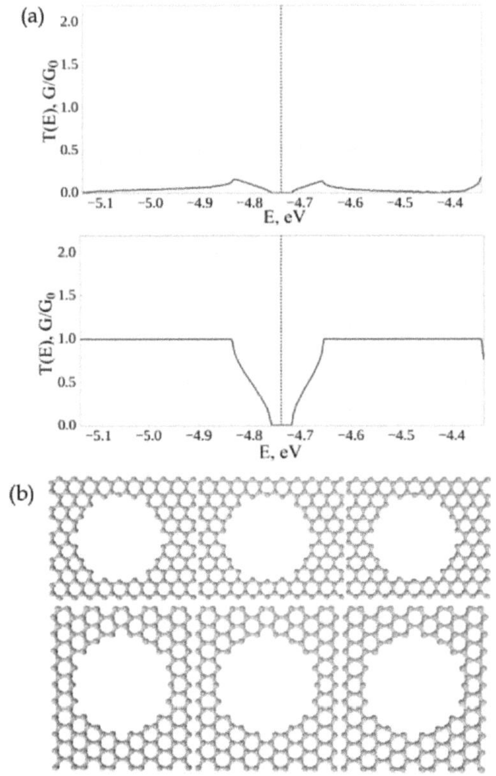

Figure 6.16 Electron transmission functions for porous graphene along the "zigzag" and "armchair" (a) directions and the atomistic models corresponding to the calculations (b).

edge. Instead of a peak of conductivity with three channels at the Fermi level, a zero value of the ETF was observed, and the width of the zero-transmission section became 0.045 eV; peaks on the sides are due to the type of ETF for graphene. In the direction along the "armchair" edge, the form of the ETF is qualitatively similar to the form of the ETF for a graphene nanoribbon of the same width. The width of the zero-transmission section decreased slightly, from 0.051 eV to 0.041 eV. The ETF region near the Fermi level, where $T(E) < 1$, increased from 0.051 eV to ~0.184 eV. The static electrical conductivities of such structures are 8.318 μS/nm and 1.497 μS/nm along the directions of the

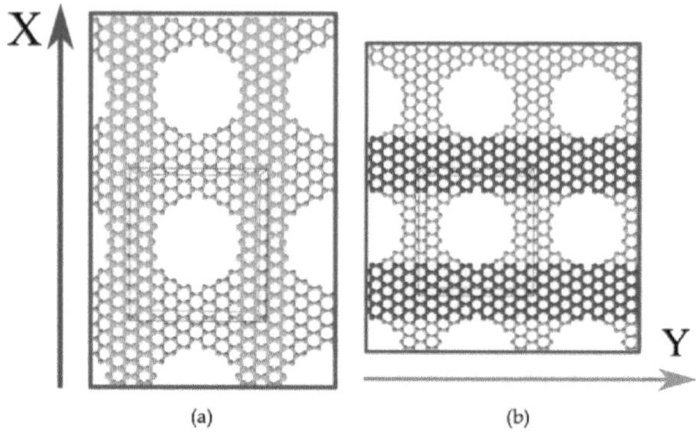

Figure 6.17 Atomistic models of porous graphene films; tracks along (a) "zigzag" edge and (b) "armchair" edge are indicated.

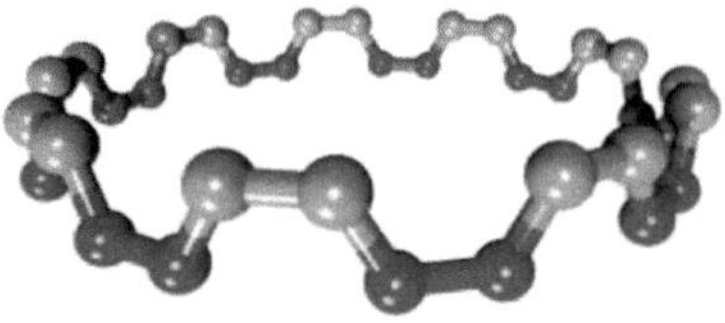

Figure 6.18 Two carbon filaments.

current transfer, "armchair" and "zigzag," respectively. The higher electrical conductivity along the "armchair" edge is due to the width of the tracks in a given direction of current transfer; along this direction, there are tracks with a width of four weakly deformed hexagons, and along the "zigzag" direction, there are three hexagons, the two extreme ones being deformed (Fig. 6.17).

4. Next, a numerical experiment was conducted to gradually increase the CNT in the pore of the graphene sheet. The buildup was performed by the layer-wise addition of carbon filaments that make up CNTs (Fig. 6.18).

First, a junction region is formed between the CNT and the graphene sheet, which contains defects in the form of pentagons, heptagons, and octagons. In Fig. 6.19, atomistic models with two carbon filaments in the pore are presented.

The form of the transmission function near the Fermi level did not change qualitatively for both directions of current transfer.

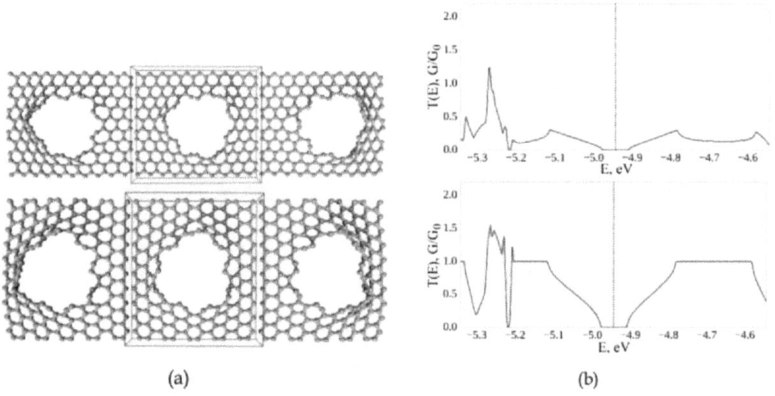

Figure 6.19 Atomistic models of porous graphene with (a) two carbon filaments and (b) ETF plots for them.

The observed difference in the region [−5.4, −5.2] would have an insignificant effect on the value of the static electrical conductivity of the material, which is 3.609 μS/nm along the "armchair" direction and 1.327 μS/nm along the "zigzag" edge for porous graphene with two layers of carbon filaments. The width of the zero-conductivity region is 0.069 eV for both directions.

When adding the next filament, we get the first layer of CNT (Fig. 6.20).

In comparison with the previous results, a sharp shift in the Fermi level in the direction of zero is observed for this structure. For this situation, the value of the Fermi energy is −4.709 eV, while for a structure with two carbon filaments, this value is −4.943 eV. The zero-conductivity region observed in the preceding cases is located to the left of the current position of the Fermi level. The width of the zero-conductivity region decreases significantly and amounts to 0.028 for both directions. The static electrical conductivity of this structure increases many times and is 22.021 μS/nm for the direction along the "armchair" edge and 9.807 μS/nm for the direction along the "zigzag" edge.

Further layer-by-layer buildup of CNTs led to the conclusion that at some stages of construction, there is a sharp increase in the static electrical conductivity of porous graphene in both directions of current transfer. Figure 6.21 shows the obtained

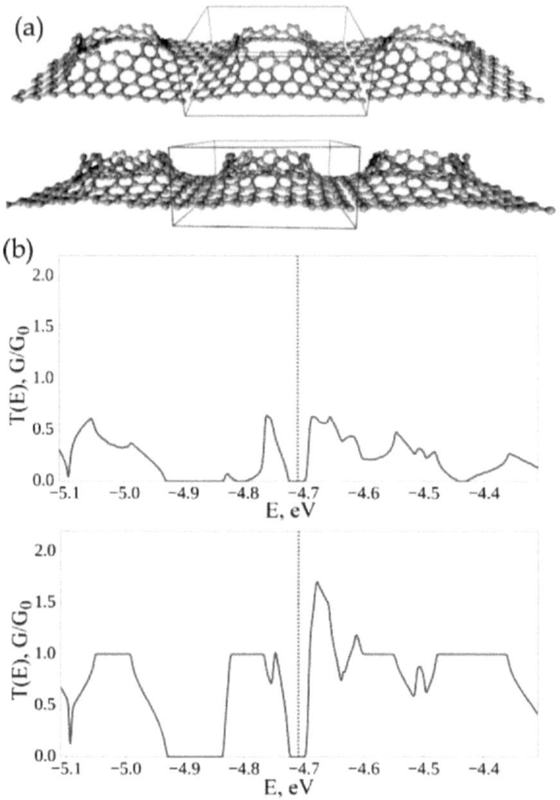

Figure 6.20 Atomistic models of porous graphene with three carbon filaments in CNT (a) and ETF plots (b) corresponding to them.

values of the static electrical conductivity as a function of the number of carbon filaments in the composition of CNTs.

This dependence directly correlates with the position of the Fermi level of porous graphene (Fig. 6.22).

5. Since the shape of the graphs depends on the length of CNTs in the porous graphene, calculations of electronic spectra of CNTs of finite size with chirality (9,9) and a diameter of 1.23 nm were performed. From Fig. 6.23, the dependence of the energy of the HOMO level and the energy gap on the number of carbon filaments in the CNTs can be shown.

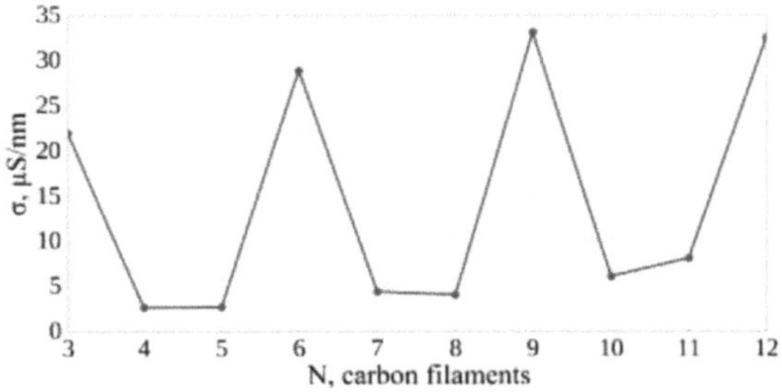

Figure 6.21 Dependence of the static electrical conductivity on the number of carbon filaments in the porous graphene.

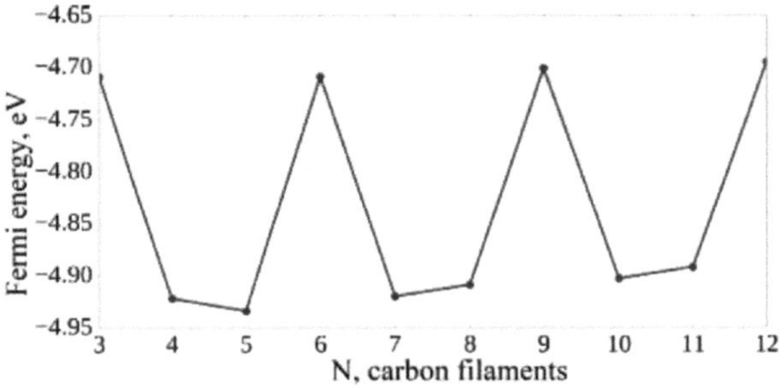

Figure 6.22 Dependence of the Fermi energy on the number of carbon filaments in the porous graphene.

It can be seen from Fig. 6.23 that sharp jumps are observed for CNTs of 3, 6, 9, and 12 carbon filaments (54, 108, 162, and 216 atoms, respectively). This fact was studied in detail by the authors of Refs. [18, 19] and was explained by the reason that CNTs of finite size form three different types of connections: the Kekule network (for example, in Fig. 6.23, this structure corresponds to structures of 4, 7, 10 filaments), incomplete Clar network (5, 8, 11 filaments), and a complete Clar network (3, 6, 9, 12 strands). These three types of connections are determined by the number of entire sections, consisting of three filaments, in the CNTs. A similar dependence of the given characteristics was observed for CNTs (5, 5) in Ref. [19].

Figure 6.24 shows the graph of the ETF along the edge of the "armchair" type of porous graphene with CNTs (9,9), consisting of six carbon filaments.

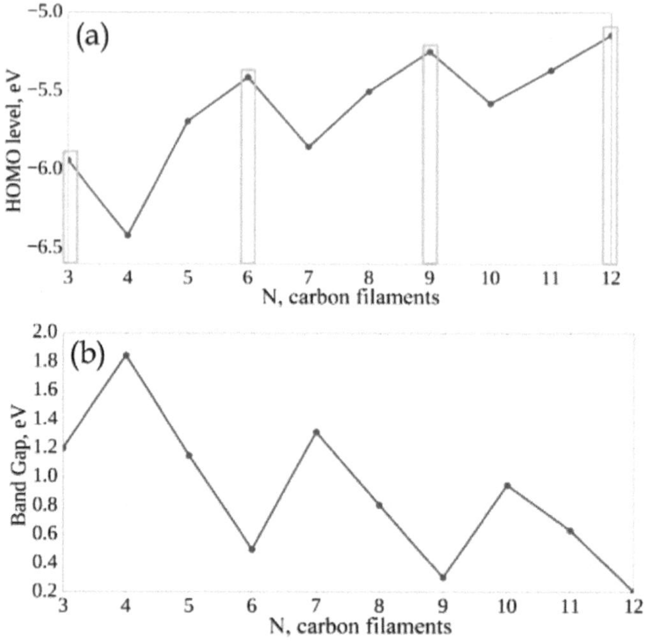

Figure 6.23 Dependence of the position of the HOMO level (a) and the energy gap value (b) on the number of carbon filaments in the carbon nanotube (9,9).

The Fermi level shifts, the ETF value on it is greater than zero, and the width of the zero-conductivity region is equal to zero.

For all the investigated CNT lengths in the porous graphene composition, a higher value of electrical conductivity along the edge of type "armchair" was observed. The results obtained for CNT and porous graphene were summarized in Table 6.1.

From the results obtained, it can be concluded that the higher electrical conductivity along the "armchair" edge is due to the geometry of the porous graphene.

In the graphs of the ETF of pillared graphene, there are clearly marked steps with a conductivity value of two. This value is due to the fact that the composition of the supercell of pillared graphene includes two segments of porous graphene.

The obtained values of the static electrical conductivity of porous graphene make it possible to conclude that for pillared graphene,

Table 6.1 Dependence of pillared graphene properties from CNT length on its composition

Carbon filaments number	Fermi energy (eV)	Static electrical conductivity (μS/nm)		Porous graphene band gap (eV)		Band gap of CNT (eV)
		"armchair"	"zigzag"	"armchair"	"zigzag"	
3	−4.709	22.021	9.807	0.028	0.028	1.2
4	−4.922	2.651	0.732	0.111	0.112	1.844
5	−4.934	2.725	0.755	0.109	0.112	1.15
6	−4.709	28.956	8.632	0.0	0.0	0.498
7	−4.92	4.411	1.244	0.092	0.095	1.312
8	−4.909	4.082	0.915	0.113	0.116	0.807
9	−4.701	33.183	9.739	0.0	0.005	0.303
10	−4.903	6.110	1.717	0.075	0.076	0.944
11	−4.892	8.133	2.027	0.069	0.069	0.63
12	−4.695	32.589	10.063	0.0	0.003	0.204

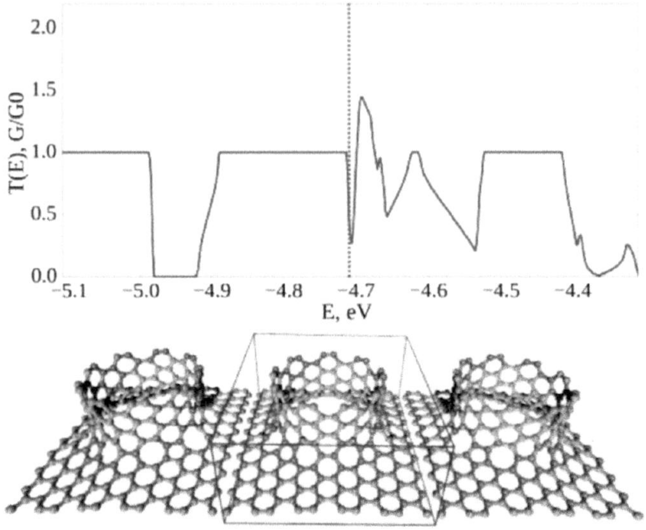

Figure 6.24 The electron transmission function for porous graphene with CNTs along the "armchair" direction and the atomistic model corresponding to the calculations (six carbon filaments).

there will be higher values of electrical conductivity at a multiple of three times the CNT length, expressed in terms of the number of carbon filaments in its composition.

For three different tube lengths, the ETF for bilayer pillared graphene films was calculated (Fig. 6.25). The dimensions of the systems studied were 4.26 nm along the y-axis and 4.8 nm along the x-axis. The distance between the CNTs within the same layer was 3.07 nm along the y-axis and 3.66 nm along the x-axis.

The ETFs were obtained using the developed method for accelerating the calculations.

From Fig. 6.26, it can be seen that after the refinement procedure, there are quite a few areas that need additional calculations. Figure 6.27 shows the final profile of the transmission function for the considered composite.

Figure 6.28 shows the transmission functions along both directions for all situations considered.

The quantity $T(E) = 3$ is due to the presence of three porous graphene sheets, and the shift in the Fermi energy in the case of the

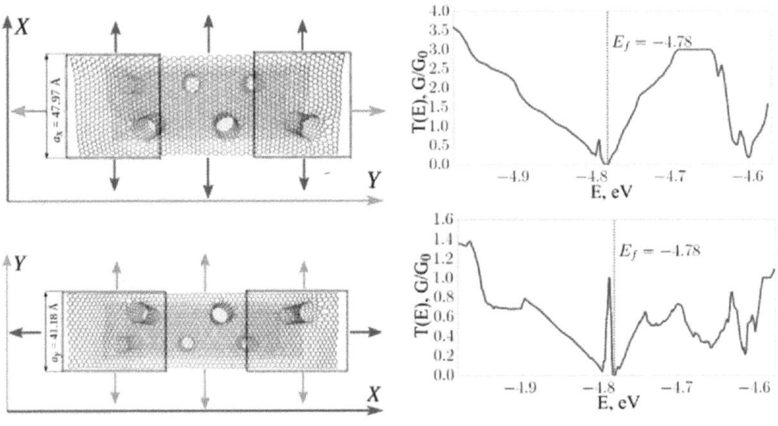

Figure 6.25 Bilayer 2D-composite pillared graphene based on CNT (9,9) 2.4 nm long: (a) atomistic models for which the quantum mechanical transparency function was calculated (electrodes and the system itself); (b) the transmission function of electrons for structures (a) along the directions *y* (left) and *x* (right).

shortest tube is observed because of the number of entire sections of three filaments in the CNT in composite.

The ETF, shown in the graphs, was used to calculate the static electrical conductivity/resistance. The transmission functions for monolayer and bilayer composites were compared. Figure 6.29 shows the graphs of transmission functions for composites with the same tube lengths $L = 1.844$ nm.

6.5 Regularities of Electrical Conductivity for Various Models of Mono- and Bilayer Films of Pillared Graphene

Using the calculated values of $T(E)$, the static electrical conductivity and the resistance of the pillared graphene films were calculated. Table 6.2 gives the necessary data for a monolayer film of pillared graphene: the length of the CNT used, the number of atoms in the supercell of the film, the value of the Fermi energy, electrical

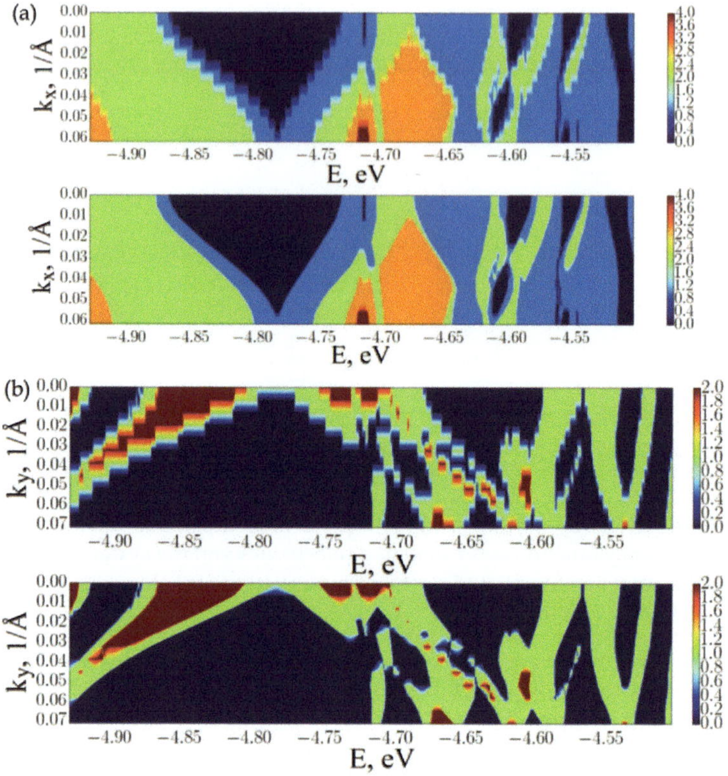

Figure 6.26 The profile of the transmission function before and after the refinement procedure for a composite containing CNTs 0.607 nm long: (a) along the *y* direction; (b) along the *x* direction.

conductivity, and resistance. The value of the Fermi energy, as can be seen from the table, is in the interval (−4.88, −4.73), that is, the level shifts compared to ideal CNTs having the same diameter (−4.69 eV). For monolayer composites in the *x* direction (axis along which the electrical conductivity was considered), the mean value of the resistance was 31.81 kΩ/nm, and in the *y* direction, it was 168.14 kΩ/nm. The maximum deviation from the average value is 43.52 kΩ/nm in the direction of infinite translation *x* and 87.99 kΩ/nm in *y*.

Similar studies were carried out for bilayer pillared graphene films (Table 6.3). The average resistance along the *x*-axis was

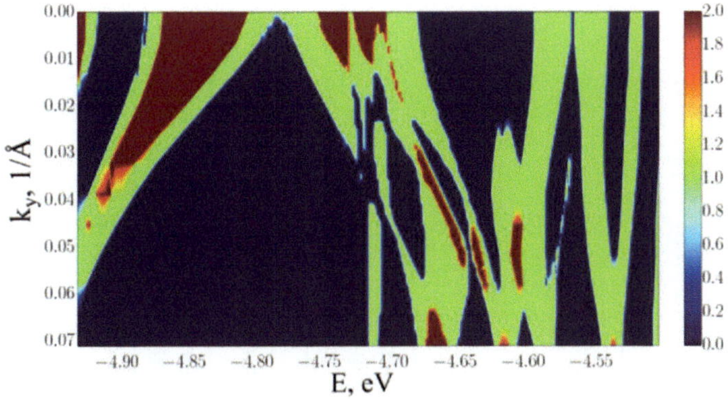

Figure 6.27 Profile of the electron transmission function for a composite with a CNT length of 0.607 nm after all refinements and additional calculations.

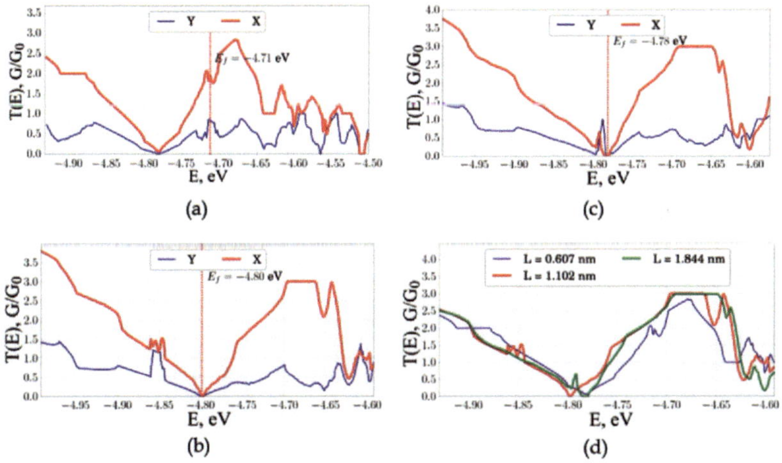

Figure 6.28 The graphs of the electron transmission function for bilayer pillared graphene films at different lengths of carbon nanotubes in its composition: (a) $L = 0.607$ nm; (b) $L = 1.102$ nm; (c) $L = 1.844$ nm; (d) comparison of the transmission functions for the three lengths examined along the *x* direction (along the edge of the "armchair" type). *x*, *y* is the direction of translation of the cell of the conducting channel.

Table 6.2 Data on the investigated supercells of monolayer composite films and the results of calculations

CNT length (nm)	Number of atoms in the supercell of composite	Translation direction	Fermi energy (eV)	Electrical conductivity (μS/nm)	Resistance (kΩ/nm)
0.60	400	*x*	−4.74	55.55	18.00
		y		5.18	193.20
0.85	436	*x*	−4.88	13.27	75.34
		y		3.90	256.13
1.10	472	*x*	−4.85	37.68	26.54
		y		11.68	85.62
1.34	508	*x*	−4.73	58.42	17.12
		y		4.89	204.71
1.59	544	*x*	−4.85	29.53	33.86
		y		6.19	161.56
1.84	580	*x*	−4.81	49.89	20.04
		y		9.29	107.62

Table 6.3 Data on the investigated supercells of bilayer composite films and the results of calculations

CNT length (nm)	Number of atoms in the supercell of composite	Translation direction	Fermi energy (eV)	Electrical conductivity (μS/nm)	Resistance (kΩ/nm)
0.60	2400	*x*	−4.712	29.45	33.96
		y		10.12	98.85
1.10	2544	*x*	−4.798	17.68	56.56
		y		7.36	135.91
1.84	2760	*x*	−4.781	17.94	55.74
		y		7.85	127.45

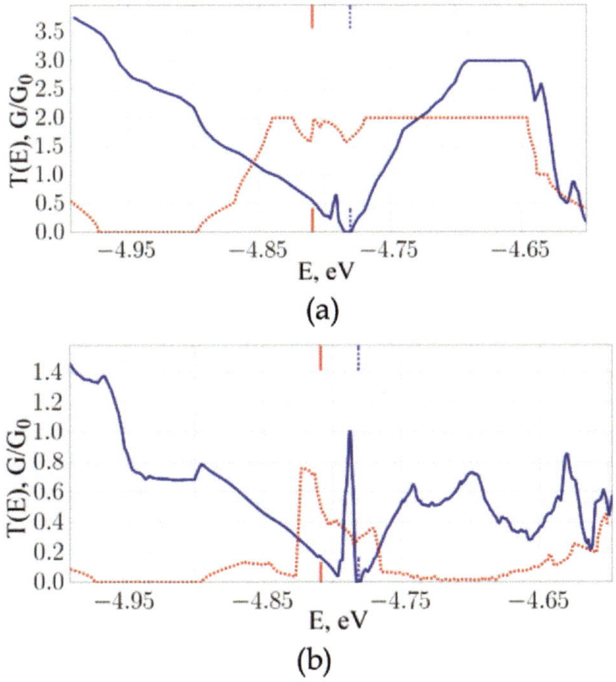

Figure 6.29 The graphs of the electron transmission functions for bilayer (blue) and monolayer (red) composites (tube length of 1.844 nm) in the direction of the *y*-axis, along the "armchair" edge (a) and the *x*-axis, along the "zigzag" edge (b). The values of the Fermi energy are marked (red segments for monolayer and blue for bilayer).

48.75 kΩ/nm, and along the *y*-axis, it was 120.74 kΩ/nm. The maximum deviation from the mean value is 24.74 kΩ/nm along the *x*-axis and 69.29 kΩ/nm along the *y*-axis. The Fermi level is shifted relative to the analogue quantity for the CNTs by the value of 0.05–0.13 eV

Figure 6.30 shows the dependence of the electrical conductivity of monolayer pillared graphene films in two directions.

From the results for monolayer and bilayer pillared graphene, the dependence of the static electric conductivity values established for porous graphene is observed. The value of electrical conductivity is always higher along the "armchair" edge and increases sharply along this direction when the composite contains CNTs, consisting

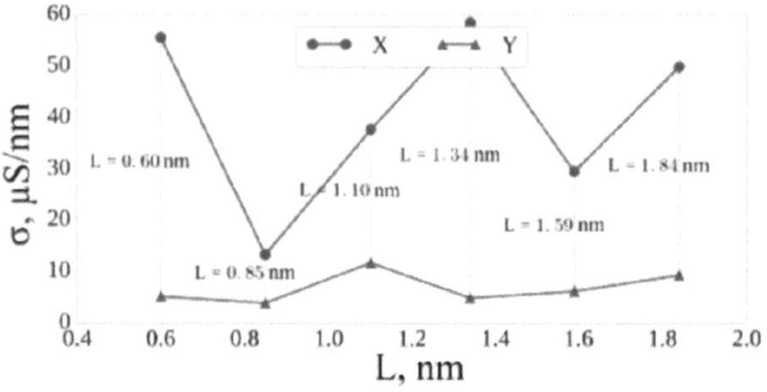

Figure 6.30 Conductivity of a monolayer film of pillared graphene depending on the distance between graphene sheets (CNT lengths).

of a multiple of three numbers of carbon filaments. In the study of current transfer in the direction of "zigzag" in a pillared graphene monolayer, a shift in the Fermi level was observed for structures with a multiple of three times the number of carbon filaments, as for porous graphene, but there was no direct dependence between this shift and the magnitude of the static electrical conductivity.

6.6 Conclusion

The energy stability of CNT–graphene composites (pillared graphene) was studied by numerical simulation methods. The heat of formation was calculated for composites, consisting of two graphene layers and a CNT (9,9) between them, attached to the sheets. All structures have a negative heat of formation and are energetically stable. From the results of calculations, it was found out that the defects are located axially symmetrically in the region of the CNT junction and the graphene sheet in the pillared graphene. This fact corresponds to the already known arrangement of carbon atoms at the junction of the end of the CNT and the hole in the graphene sheet [5, 12, 16]. The most stable of those considered are systems in which the diameter of CNTs is the largest (1.23 nm), which can be explained by the number of bonds between the CNT and the graphene sheet.

The calculated electrical conductivity values of pillared graphene films are rather low for the considered situations because of the size of the composite supercells. Normalization was carried out on the sample size in the corresponding direction in nanometers. To convert electrical conductivity to a parameter of a large cell with the dimension S/m, there is a restriction associated with the incorrectness of the direct transition. For large dimensions of the cells under consideration, it is necessary to take into account the size effect and the mean free path of the electron. Therefore, a direct conversion of the conductivity for a single-layer film leads to incorrect values of the order of ~50 kS/m, which, however, allow us to suggest a high conductivity of the films with a larger size of the samples.

The calculated characteristics of the pillared graphene films—electrical conductivity and resistance in the "armchair" and "zigzag" directions of the graphene sheet—show that the resistance along the "armchair" edge for both types of composites (monolayer and bilayer) is lower than in the "zigzag" direction. The magnitude of the electrical conductivity in a given direction depends on the number of carbon filaments of the CNTs in the composite. The average resistance value for a monolayer composite is 31.81 kΩ/nm. For a bilayer composite, the average resistance is 48.75–120.74 kΩ/nm, depending on the direction. The high electrical conductivity along the "armchair" edge is due to the view of the ETF of the porous graphene, which in turn is determined by the width of the graphene tracks of this type between the pores: four hexagons constitute a path along the "chair" direction. For the "zigzag" direction, three hexagons constitute a path, of which two are significantly deformed.

In the "armchair" direction of the graphene sheet, the electrical conductivity changes with a larger amplitude, which is two–three times higher in absolute value than the electrical conductivity for the direction along the "zigzag" type edge—the resistance varies greatly with the increase in the length of CNTs.

It should be noted that the increased resistance along the "zigzag" edge and the reduced resistance along the "armchair" edge show more smoothed conductive characteristics of the bilayer composite, depending on the direction of current transfer.

In general, an important conclusion can be made: Films of pillared graphene with nanotubes 1.23 nm in diameter are characterized by a relatively high electrical conductivity, and their characteristics can be controlled through the length of CNTs in the composite. Due to their high mechanical strength and electrical conductivity, they can be successfully applied in electronic devices and as electrodes of storage batteries. Besides, their structure contains a lot of pores and cavities, which allows filling with various necessary compounds.

References

1. Labunov, V. A., Shulitski, B. G., Prudnikava, A. L., Shaman, Y. P., and Basaev, A. S. (2010). Composite nanostructure of vertically aligned carbon nanotube array and planar graphite layer obtained by the injection CVD method, *Semiconductor Phys. Quantum Electron. Optoelectron.*, **13(2)**, pp. 137–141.
2. Du, F., Yu, D., Dai, L., Ganguli, S., Varshney, V., and Roy, A. K. (2011). Preparation of tunable 3D pillared carbon nanotube graphene networks for high-performance capacitance, *Chem. Mater.*, **23**, pp. 4810–4816.
3. Kondo, D., Sato, Sh., and Awano, Y. (2011). Self-organization of novel carbon composite structure: Graphene multi-layers combined perpendicularly with aligned carbon nanotubes, *Appl. Phys. Exp.*, **1**, 074003.
4. Das, S., Seelaboyina, R., Verma, V., Lahiri, I., Hwang, J. Y., Banerjee, R., and Choi, W. (2011). Synthesis and characterization of self-organized multilayered graphene–carbon nanotube hybrid films, *J. Mater. Chem.*, **21**, 7289.
5. Matsumoto, T. and Saito, S. (2002). Geometric and electronic structure of new carbon-network materials: Nanotube array on graphite sheet, *J. Phys. Soc. Jpn.*, **71(11)**, pp. 2765–2770.
6. Wesołowski, R. P. and Terzyk, A. P. (2016). Dynamics of effusive and diffusive gas separation on pillared grapheme, *Phys. Chem. Chem. Phys.*, **18**, pp. 17018–17023.
7. Loh, G. C., Teo, E. H. T., and Tay, B. K. (2011). Interpillar phononics in pillared-graphene hybrid nanostructures, *J. Appl. Phys.*, **110**, 083502.
8. Loh, G. C., Teo, E. H. T., and Tay, B. K. (2012). Tuning the Kapitza resistance in pillared-graphene nanostructures, *J. Appl. Phys.*, **111**, 013515.
9. Yang, Y., Kim, N. D., Varshney, V., Sihn, S., Li, Y., Roy, A. K., Tour, J. M., and Lou, J. (2017). In situ mechanical investigation of carbon

nanotube–graphene junction in three-dimensional carbon nanostructures, *Nanoscale*, **9**, pp. 2916–2924.

10. Barako, M. T., Roy-Panzer, S., English, T. S., Kodama, T., Asheghi, M., Kenny, T. W., and Goodson, K. E. (2015). Thermal conduction in vertically aligned copper nanowire arrays and composites, *ACS Appl. Mat. Int.*, **7(34)**, pp. 19251–19259.
11. Lin, J., Zhong, J., Bao, D., Reiber-Kyle, J., Wang, W., Vullev, V., Ozkan, M., and Ozkan, C. S. (2012). Supercapacitors based on pillared graphene nanostructures, *J. Nanosci. Nanotechnol.*, **12(3)**, pp. 1770–1775.
12. Shayeganfar, F. and Shahsavari, R. (2016). Electronic and pseudomagnetic properties of hybrid carbon/boron nitride nanomaterials via ab-initio calculations and elasticity theory, *Carbon*, **99**, pp. 523–532.
13. Brenner, D. W. (1990). Empirical potential for hydrocarbons for use in simulating the chemical vapor deposition of diamond films, *Phys. Rev. B*, **42(15)**, pp. 9458–9471.
14. Stuart, S. J., Tutein, A. B., and Harrison, J. A. (2000). A reactive potential for hydrocarbons with intermolecular interactions, *J. Chem. Phys.*, **112(6472)**, pp. 760–767.
15. Glukhova, O. E., Kolesnikova, A. S., Slepchenkov, M. M., and Shmygin, D. S. (2015). Atomic structure of energetically stable carbon nanotubes/graphene composites, *Phys. Solid State*, **57(5)**, pp. 1009–1013.
16. Dimitrakakis, G. K., Tylianakis, E., and Froudakis, G. E. (2008). Pillared graphene: A new 3-D network nanostructure for enhanced hydrogen storage, *Nano Lett.*, **8(10)**, pp. 3166–3170.
17. Elstner, M., Porezag, D., Jungnickel, G., Elsner, J., Haugk, M., Frauenheim, T., Suhai, S., and Seifert, G. (1998). Self-consistent-charge density-functional tight-binding method for simulations of complex materials properties, *Phys. Rev. B*, **58**, 7260.
18. Dou, K. P., Fu, X., De Sarkar, A., and Zhang, R. (2015). Tailoring the transmission lineshape spectrum of zigzag graphene nanoribbon based heterojunctions via controlling its width and edge protrusions, *Nanoscale*, **7**, pp. 20003–20008.
19. Matsuo, Y., Tahara, K., and Nakamura, E. (2003). Theoretical studies on structures and aromaticity of finite-length armchair carbon nanotubes, *Organic Lett.*, **5(18)**, pp. 3181–3184.

Chapter 7

Electro-conductive Properties and Applications of Graphene-DNA Nucleotide Complex

Dmitry Shmygin[a] **and Olga E. Glukhova**[a,b]

[a]*Physical Department, Saratov State University, 83 Astrakhanskaya Street 410010, Saratov, Russia*

[b]*I. M. Sechenov First Moscow State Medical University, Laboratory of Biomedical Nanotechnology, Trubetskaya Street 8-2, Moscow, 119991, Russia*

glukhovaoe@info.sgu.ru

The electronic properties of graphene nanoribbon segments in the presence of DNA nucleotides were studied in a series of numerical experiments. Segments with both armchair and zigzag edges were considered. These segments were bonded to semi-infinite graphene sheets and graphene nanoribbons. The study was carried out to identify the possibility of using a given structure as a biosensor, due to which it is possible to separate one biological molecule from another, in particular, DNA nucleotides.

7.1 Introduction

Graphene is a monolayer of carbon atoms that are located at the nodes of the hexagonal lattice. The material has unique properties,

2D and 3D Graphene Nanocomposites: Fundamentals, Design, and Devices
Edited by Olga E. Glukhova

ISBN 978-981-4800-41-9 (Hardcover), 978-0-429-20150-9 (eBook)
www.jennystanford.com

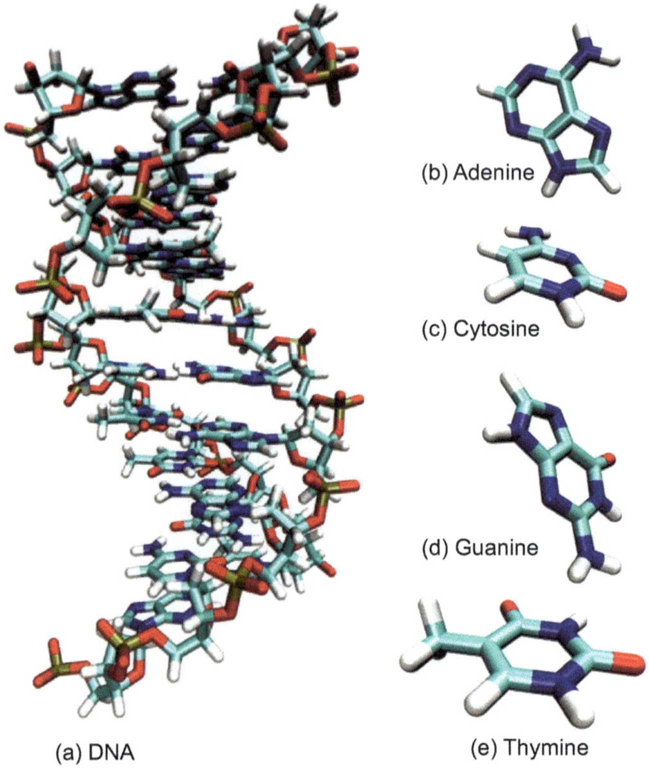

Figure 7.1 Atomistic DNA model and four nucleotides.

such as high mechanical strength [1, 2], thermal conductivity [3], and electrical conductivity [4], which allowed its application in many areas of electronics [5–7]. DNA sequencing is an important task, and the search for more convenient, cheaper, faster, and more accurate ways of its solution continues. One of the possible ways to solve this problem is to use graphene and graphene nanoribbons, whose conductivity, possibly, varies depending on the presence of DNA nucleotides (Fig. 7.1) nearby. Several works have talked about the prospects of porous graphene nanoribbons for solving the problem of DNA sequencing. In this chapter, we theoretically investigate the effect of single nucleotides on the transmission function and conductivity of graphene nanoribbons without nanopores.

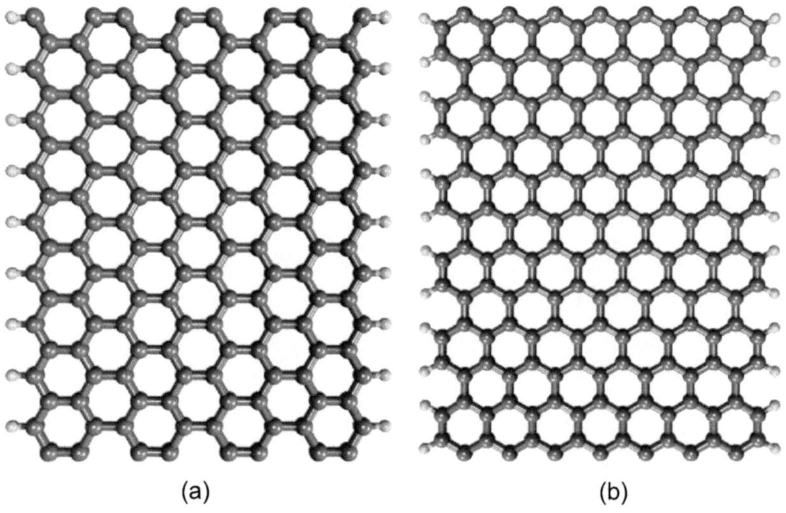

Figure 7.2 (a) A segment of a nanoribbon with a hydrogenated zigzag edge: length ~2.1 nm, width ~1.8 nm; (b) a segment of nanoribbon with a hydrogenated armchair edge: length ~2.4 nm, width ~1.9 nm (for visualization, the graphic part KVAZAR [8]) was used.

Despite the similarity of geometry, armchair and zigzag graphene nanoribbons (Fig. 7.2) significantly differ in their electronic properties. For this reason, ribbons of both types were considered as substrate segments contacting nucleotides (Fig. 7.3).

The edges of the graphene sheets were hydrogenated to avoid edge effects. In this situation, periodic boundary conditions were not used. For substrate-supported nanoribbon of another type, periodic boundary conditions were applied in one direction.

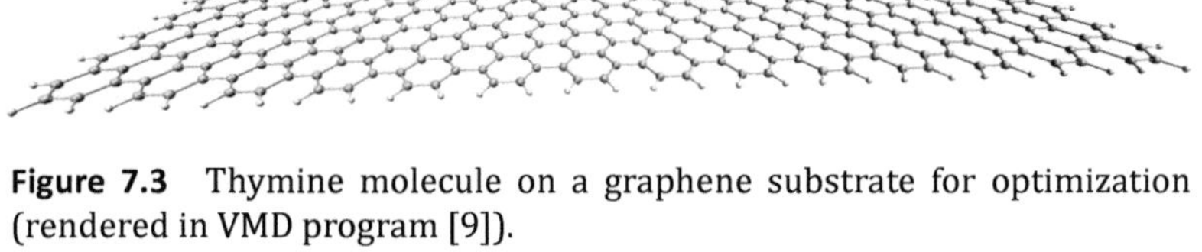

Figure 7.3 Thymine molecule on a graphene substrate for optimization (rendered in VMD program [9]).

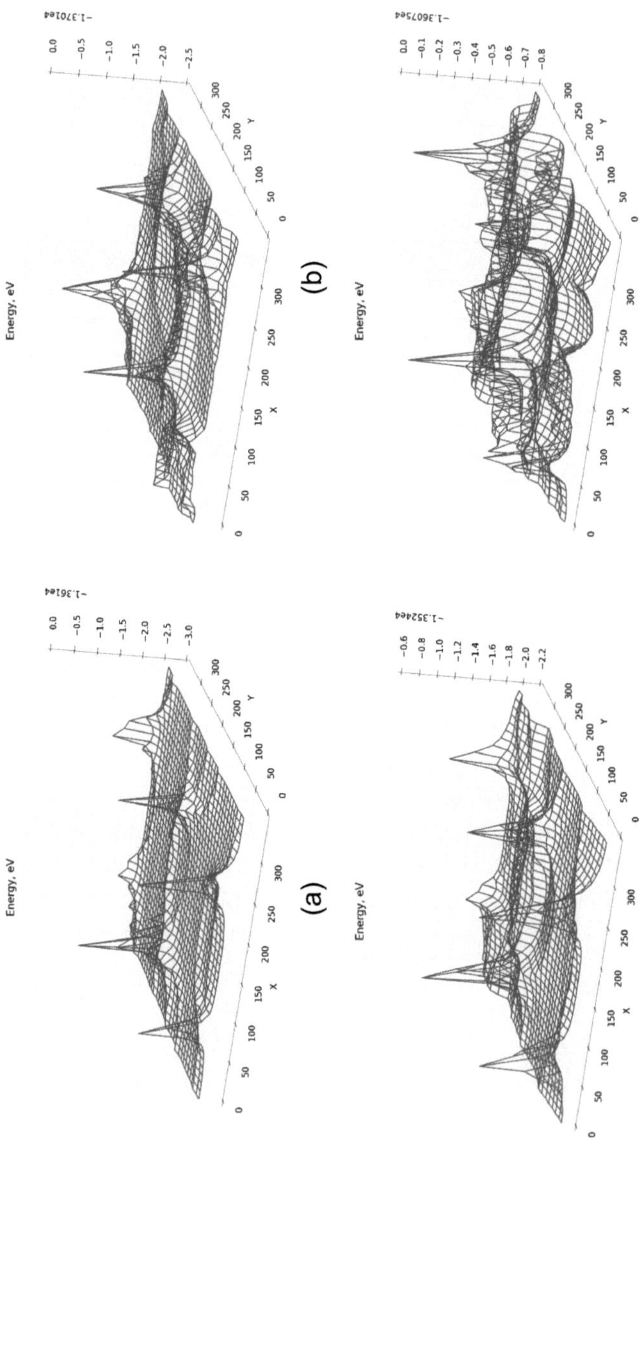

Figure 7.4 Dependence of potential energy of the system with different nucleotides (a, adenine; b, guanine; c, cytosine; d, thymine) from current rotation angle (starts from the initial position) around x- and y-axes.

7.2 Method

The behavior of the presented systems was modeled by the density functional method in the tight-binding approximation with self-consistent charge calculation (SCC DFTB) [10–12]. Prior to the calculation of the transmission function, the geometry of the presented structure was optimized. Optimization of this kind will lead the system to a state characterized by the potential energy minima (conjugate-gradient method was used). The position of the nucleotide above the surface of the graphene sheet can be more optimal with a more successful starting configuration. To test and use this assumption, a number of static calculations were performed, in each of which the nucleotide was rotated relative to the substrate. Based on the obtained values, three-dimensional figures of the dependence of the potential energy on the current angles of rotation were constructed (Fig. 7.4).

On the basis of these graphs, the position of the nucleotide with the lowest energy was chosen. Then it was used as the starting point of the optimization procedure.

To calculate the electrical conductance, we use the Green–Keldysh functions and the Landauer–Büttiker formalism. The conductivity of the materials was calculated with the following expression:

$$G = 2\frac{e^2}{h}\int T\left(E\right) F_{\mathrm{T}}\left(E\right) dE, \tag{7.1}$$

where

$$F_{\mathrm{T}}\left(E\right) = \frac{1}{4K_{\mathrm{B}}T}\sec h^2\left(\frac{E-\mu}{2K_{\mathrm{B}}T}\right),$$

$T(E)$ is the transmission function, E is the energy value, μ is the Fermi energy of the contacts, T is the temperature, K_{B} is the Boltzmann constant, e is the electron charge, and h is the Planck's constant. The resistance values were obtained from the calculated conductivity. The transmission function can be found according to the next equation:

$$T\left(E\right) = \frac{1}{N} \times \sum_{k=1}^{N} \mathrm{Tr}\left(\Gamma_{\mathrm{S}}\left(E\right) G_{\mathrm{C}}^{\mathrm{A}}\left(E\right) \Gamma_{\mathrm{D}}\left(E\right) G_{\mathrm{C}}^{\mathrm{R}}\left(E\right)\right), \tag{7.2}$$

where $G_C^A(E)$, $G_C^R(E)$ are the advanced and retarded Green matrices, respectively (describing the contact with the electrodes); $\Gamma_S(E)$, $\Gamma_D(E)$ are the level broadening matrices for the source and drain. The averaged transmission function $T(E)$ can be found by the summation of the transmission functions over the entire Brillouin zone (BZ) and N is the number of points in the reciprocal space. For one-dimensional systems, the calculation is made at one point, and for two-dimensional systems, all the points of the reciprocal space lie on one line.

7.3 Results and Discussion

The difference in energy between the worst and optimal position is given in Table 7.1 for each nucleotide.

Table 7.1 Potential energies of systems with nucleotides at different stages of calculations (zigzag nanoribbon serves as substrate)

Nucleotide	The worst energy (eV)	Optimal energy (eV)	Energy after optimization (eV)
Adenine	−13610.494	−13612.5786	−13655.7601
Guanine	−13701.3202	−13703.0524	−13746.2302
Cytosine	−13524.7932	−13526.0877	−13569.2563
Thymine	−13607.5371	−13608.2799	−13651.4430

All the calculations were made using the self-consistency charge procedure (Fig. 7.5).

Models with two different electrodes were considered. For zigzag-type nanoribbons, segments of the same nanoribbon were used as electrodes, which were transmitted to infinity along the direction of current propagation. For nanoribbons of type armchair, semi-infinite two-dimensional electrodes oriented with the corresponding side to the substrate were used. Corresponding sizes are shown in Fig. 7.6.

The transmission of one pure zigzag nanoribbon model can be found in Fig. 7.7.

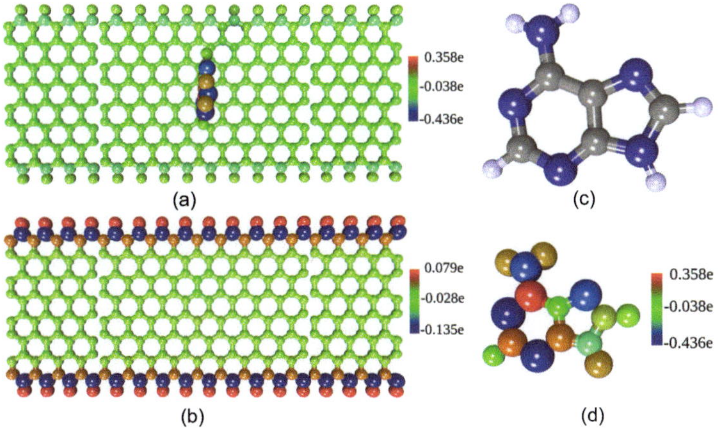

Figure 7.5 Adenine molecule above the surface of a graphene sheet: (a) charge distribution with adenine and substrate; (b) charge distribution for substrate only; (c) adenine molecule; (d) charge distribution for adenine molecule only.

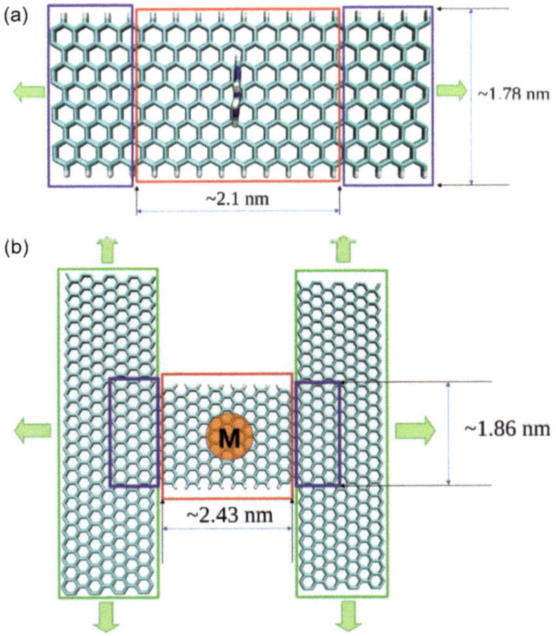

Figure 7.6 Schemes for all transmission calculations: (a) one-dimensional system for zigzag nanoribbons; (b) two-dimensional semi-infinite electrodes with the armchair nanoribbon as the substrate.

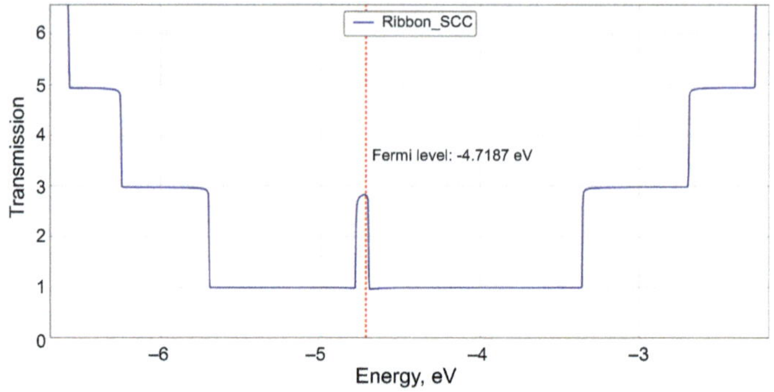

Figure 7.7 Transmission function for pure carbon zigzag nanoribbon (more detailed calculations).

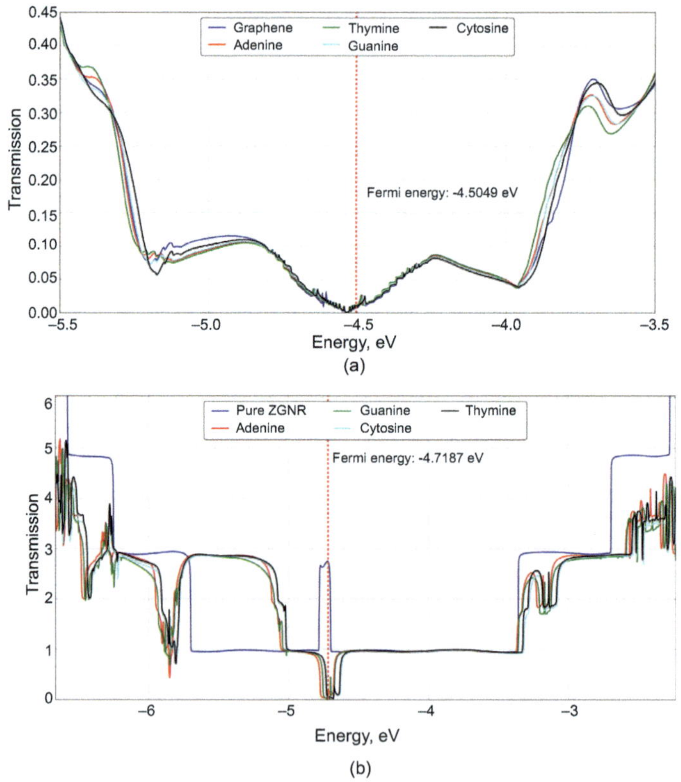

Figure 7.8 The transmission function of the considered structure in which the substrate is (a) the armchair nanoribbon segment and (b) a segment of zigzag nanoribbon.

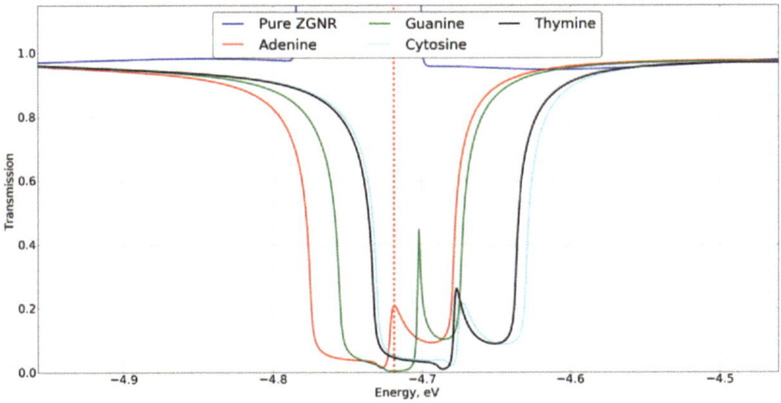

Figure 7.9 The transmission function of the considered structure near the Fermi energy, where the substrate is a segment of a zigzag nanoribbon.

The program package KVAZAR [8] was used to make numerical simulations. Figure 7.8 shows the transmission function for armchair and zigzag segments of graphene nanoribbons with different single nucleotides.

There are various peaks of the transmission function near the Fermi energy for the different nucleotides (Fig. 7.9).

The armchair nanoribbon is characterized by low values of the transmission function near the Fermi energy that leads to a very high resistance value for the substrate structure. The presence of nucleotides had no significant effect on it. Zigzag nanoribbon is a conductor, and the presence of nucleotides increased the resistance of the segment. In the presence of adenine, thymine, guanine, and cytosine, the resistance increased accordingly by 6.9, 5.5, 6.2, and 5.3 times and equaled to 45.49 kΩ, 36.25 kΩ, 40.99 kΩ, and 35.02 kΩ.

By using the resistance value of the zigzag segment with the nucleotide, it is possible to determine which nucleotide is located on the surface of the segment of the graphene nanoribbon. This makes it possible to use such systems as biosensors, particularly for DNA nucleotides.

Acknowledgment

The "Russian Foundation For Basic Research" supported this work (project 15-29-01025).

References

1. Lee, C., Wei, X., Kysar, J. W., and Hone, J. (2008). Measurement of the elastic properties and intrinsic strength of monolayer graphene, *Science*, **321(5887)**, pp. 385–838.
2. Memariana, F., Fereidoona, A., and Ganjib, M. D. (2015). Graphene Young's modulus: Molecular mechanics and DFT treatments, *Superlattices Microstruct.*, **85**, pp. 348–356.
3. Pop, E., Varshney, V., and Roy, A. K. (2012). Thermal properties of graphene: Fundamentals and applications, *MRS Bull.*, **37(12)**, pp. 1273–1281.
4. Neto, A. C., Peres, N. M. R., Novoselov, K. S., and Geim, A. K. (2009). The electronic properties of graphene, *Rev. Mod. Phys.*, **81**, pp. 109–162.
5. Sagade, A. A., Neumaier, D., Schall, D., Otto, M., Pesquera, A., Centeno, A., Elorza, A. Z., and Kurz, H. (2015). Highly air stable passivation of graphene based field effect devices, *Nanoscale*, **7**, pp. 3558–3564.
6. Avouris, P., Chen, Z., and Perebeinos, V. (2007). Carbon-based electronics, *Nat. Nanotechnol.*, **2(10)**, pp. 605–615.
7. Chee, W. K., Lim, H. N., Zainal, Z., Huang, N. M., Harrison, I., and Andou, Y. (2016). Flexible graphene-based supercapacitors: A review, *J. Phys. Chem. C.*, **120(8)**, pp. 4153–4172.
8. http://nanokvazar.ru/ (2 January 2018).
9. Humphrey, W., Dalke, A., and Schulten, K. (1996). VMD: Visual molecular dynamics, *J. Molec. Graphics*, **14**, pp. 33–38. http://www.ks.uiuc.edu/Research/vmd/.
10. Elstner, M., Porezag, D., Jungnickel, G., Elsner, J., Haugk, M., Frauenheim, Th., Suhai, S., and Seifert, G. (1998). Self-consistent-charge density-functional tight-binding method for simulations of complex materials properties, *Phys. Rev. B*, **58**, pp. 7260–7268.
11. Pecchia, A., Penazzi, G., Salvucci, L., and Di Carlo, A. (2008). Non-equilibrium Greens functions in density functional tight binding: Method and applications, *New J. Phys.*, **10**, 065022.
12. Papaconstantopoulos, D. A. and Mehl, M. J. (2003). The Slater–Koster tight-binding method: A computationally efficient and accurate approach, *J. Phys. Condens. Matter*, **15**, pp. 413–440.

Chapter 8

Structural and Electronic Properties of a New Multilayer Graphene/Graphane Material

Olga E. Glukhova,[a,b] Vladislav V. Shunaev,[a] and Michael M. Slepchenkov[a]

[a]*Physical Department, Saratov State University, 83 Astrakhanskaya Street 410010, Saratov, Russia*
[b]*I. M. Sechenov First Moscow State Medical University, Laboratory of Biomedical Nanotechnology, Trubetskaya Street 8-2, Moscow, 119991, Russia*
glukhovaoe@info.sgu.ru

This chapter is devoted to the study of a new layered hybrid structure based on graphene and graphane. Several types of stacking of the resulting composite are considered. We calculated the band structure and the density of electronic states (DOS) of the investigated composite using the quantum density-functional-based tight-binding (DFTB) method. From the analysis of the calculated data, the prospects of the proposed hybrid graphene/graphane structure as an element base of radio electronic devices are predicted.

8.1 Introduction

Graphene demonstrates unique mechanical [1], electronic, and magnetic properties that can be varied by size and chirality [2, 3].

2D and 3D Graphene Nanocomposites: Fundamentals, Design, and Devices
Edited by Olga E. Glukhova

ISBN 978-981-4800-41-9 (Hardcover), 978-0-429-20150-9 (eBook)
www.jennystanford.com

Hydrogenated graphene, also known as graphane, has an energy gap that lies in the range of 3.5–4.5 eV [4–7]. Recently, a lot of attention has been given to hybrid graphene/graphane nanocomposites since their combination can find application in thin, flexible displays, solar panels, and waveguides [8]. Hybrid structures based on graphene and graphane are synthesized from graphite in a medium of hydrogen gas at a pressure of more than 2 GPa and at a temperature of 450–700°C [9]. Another way of forming such structures is laser irradiation of graphene, on the surface of which H_2O or NH_3 molecules are adsorbed [10]. In the latter case, hydrogenation of graphene is possible on only one side of the sheet.

Methods of mathematical modeling are helpful for studying properties of graphene/graphane nanostructures. Reddy et al. [11] studied semi-zigzag, full-zigzag, and full-armchair graphene and graphane interfaces by finite-element and atomistic simulations. Values of mismatch strain, tilt angle, and interfacial stress were calculated by the adaptive intermolecular reactive empirical bond order (AIREBO) potential. The finite element method (FEM) modeling has shown that interfacial stress is not critical for mismatch strain in comparison to rippling and warping of the graphene–graphane composite sheet.

By the first-principle calculations, Jin et al. found the optical anisotropy of graphene/graphane superlattices with zigzag interfaces depending on the atomic topology [12]. The main role in optical properties of graphene/graphane superlattices is played by the graphene's width and the graphane's coplanar states of 2px, 2py. It was found that all considered superlattices demonstrated high reflective polarization along with low reflectivity and high transmissivity, which is good for application in ultraweak photon emission source.

Averill and Morris [13] showed that graphene/graphane interfaces demonstrate useful electronic properties that depend on defects. It was found that significant strains occur in both graphene and graphane regions due to the presence of an interface, with both area strains and shear strains commonly close to 1%. Interfaces energies for the zigzag are 0.12 eV/Å per unit length, and those for the armchair are 0.11 eV/Å. The area of defect alternates the electronic properties of these composite materials. By the tight-binding scheme, Schmidt and Loss studied the electronic properties of graphene/graphane heterostructures [14]. It was found that the zigzag-oriented interfaces created effective edge for the π-electrons of graphene. Such edge states lead to the presence of quantum spin and valley Hall effects. The density functional theory (DFT) calculation showed that after the Li-doping of graphene/graphane nanocomposite, most of charge is transferred to the upper graphane layer that behaves like a classical plate capacitor [15]. Analysis of the DOS for graphane/graphene nanoribbons ($AGA_{13-x}/GxNRs$) confirmed that it is all direct semiconductor and that the band structure near the Fermi level is mainly determined by graphene addition [16]. It was also shown that magnetic properties of such nanocomposite can be tuned.

In this chapter, we suggest a new hybrid structure that consists of two graphene and one graphane layer. The goal of this chapter is to study the energy parameters, as well as the electronic and conductive properties of this structure by methods of quantum chemistry.

8.2 Construction of Graphene/Graphane Atomic Model

We considered a three-layered composite where the graphane sheet was located between two graphenes. The length of all sheets was equal to 24.14 Å in the armchair direction and 17.22 Å in the zigzag direction. This chapter studies three types of layered hybrid graphene/graphane material: (1) type AAA where all layers located strictly one above the other; (2) type AAB where the upper layer is shifted in relation to the two others; (3) type ABA where the medium layer is shifted. The graphene/graphane atomic models were constructed by the KVAZAR software [17]. The unit cells of these structures are shown in Fig. 8.1.

The full structures can be received from these unit cells by translation. The obtained structures with fixed edges were optimized by minimization of the total energy calculated within the AIREBO method [18]. The AIREBO potential consists of the Brenner potential for bonded atoms, the Lennard–Jones potential for nonbonded

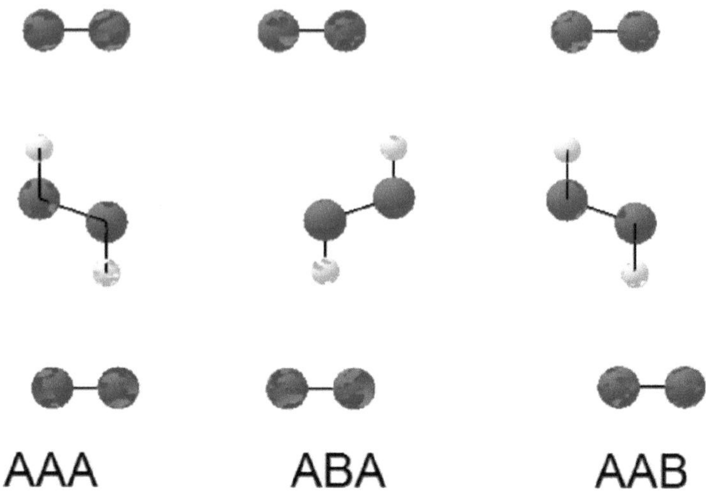

Figure 8.1 Unit cells of layered graphene/graphane nanocomposites.

atoms, and the potential for dihedral angles. This method was developed for hydrocarbon structures and allows to model the processes of chemical reactions, mechanical deformations, and stability of composite materials. In the result of optimization, ground states for graphene/graphane nanocomposites of three types were found. Their atomic structures are shown in Fig. 8.2. It can be seen that the edges of the graphene and graphane layers are bent to meet each other wrapping a graphane layer between them. At the same time, the central region of all three sheets remained practically unchanged.

The average values of energy per atom before and after optimization as well as the heat of formation for all types of packing are shown in Table 8.1. Let us note that the calculated values of heat of formation are in good agreement with the known experimental value of graphane-like structures [19]. As seen from Table 8.1, the structure of AAA type has the lowest energy and the lowest heat of formation, so it will be the most thermodynamically stable and energy favorable for synthesis. It is important that all considered nanocomposites have negative heat of formation, so they all can exist.

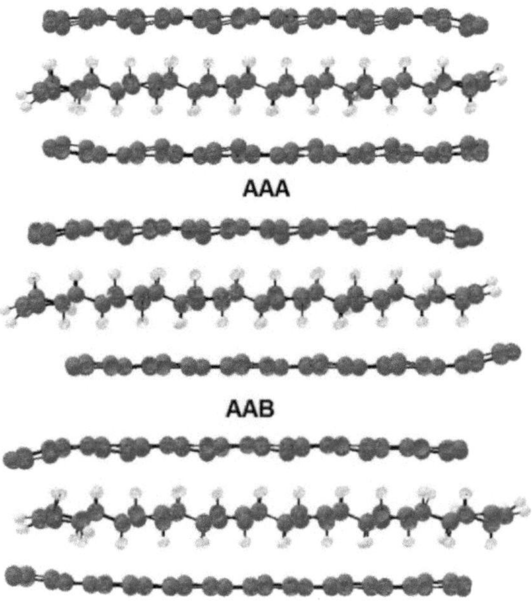

Figure 8.2 Layered nanocomposite graphene/graphane with different types of packing: AAA, AAB, ABA after optimization by the AIREBO method.

Table 8.1 Total energy of graphene/graphane nanocomposites before and after optimization

Type of packing	Total energy before optimization (eV/atom)	Total energy after optimization (eV/atom)	Heat of formation (eV/atom)
AAA	−5.868	−4.453	−0.025
ABA	−5.864	−4.925	−0.007
AAB	−5.867	−4.711	−0.023

8.3 Electronic Properties

At the next stage, we estimated the electronic properties of the obtained structures. To find the electric structure of graphene/graphane material, we used the SCC DFTB 2 [20] method with the

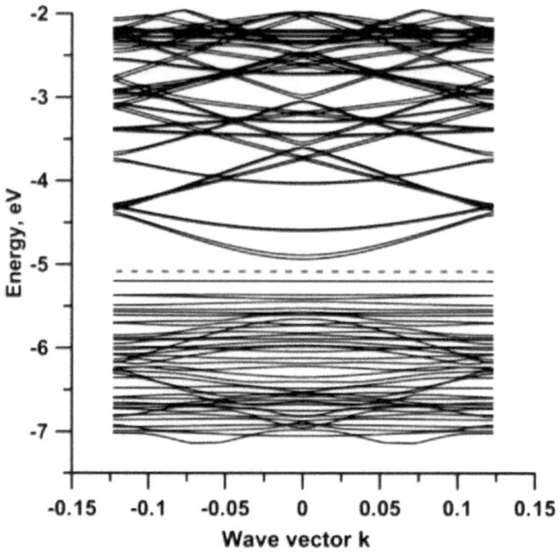

Figure 8.3 Zone structure of the graphene/graphane composite AAA.

parameter set "3ob-3-1". The zone structure of composite AAA is shown in Fig. 8.3 (dashed line marks the Fermi level). For this structure, the Fermi level equals −5.1 eV. The rise in Fermi level in comparison to pure graphane ($E_f = -5$ eV) indicates a downgrade of emissive property due to a decrease in the work function. The band gap between the conductive zone (CZ) and the valence zone (VZ) is clearly seen and composes about 0.55 eV. It can be concluded that the composite AAA has a semiconductive type of conductivity in contrast to dielectric graphane and conductive graphene. Figure 8.3 shows that near the Fermi level, the VZ of AAA composite is speckled by lines, which can be explained by the contribution of graphane, while the CZ is almost similar to that of graphene.

The electric structures of composites ABA and AAB differ from the structure of AAA (Fig. 8.4). First of all, they have much bigger different Fermi level: −5.27 for ABA and −5.28 eV for AAB. So these two composites would have the worst properties than the AAA structure. The composite ABA demonstrates the biggest gap of 0.65 eV, while for the composite AAB, it practically equals to the band gap of the AAA structure (0.53 eV). It can be explained by

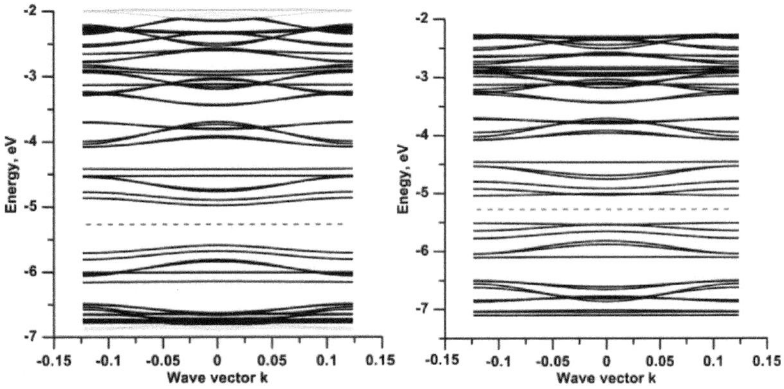

(a) (b)

Figure 8.4 Zone structure of graphene/graphane composites: (a) ABA type; (b) AAB type.

the topologies of these structures and by their relative packing that influence the location of electron clouds.

On the basis of the calculated zone structures, the DOS spectra were obtained for all structures (dashed lines correspond to the position of Fermi level) (Fig. 8.5). All the graphs confirm that graphene/graphane composites have semiconductive type of conductivity. The peak near the Fermi level for all structures typical for armchair graphene can be clearly seen. The graphs clearly show high DOS for all types of composites, which indicate the prospects of these materials in different applications.

8.4 Conductive Properties

To estimate the conductive properties of the obtained graphene/graphane composites, we used the non-equilibrium Green's function (NEGF) method described in Chapter 3. We used graphene nanoribbons of one hexagon length as left and right electrodes (Fig. 8.6). The difference in topology between the central molecule and the electrode may affect the conductive properties of the considered composite. The transmission function for the composite of AAA type is shown in Fig. 8.6 (units of measurement $G_0 = e^2/h$). It can be seen

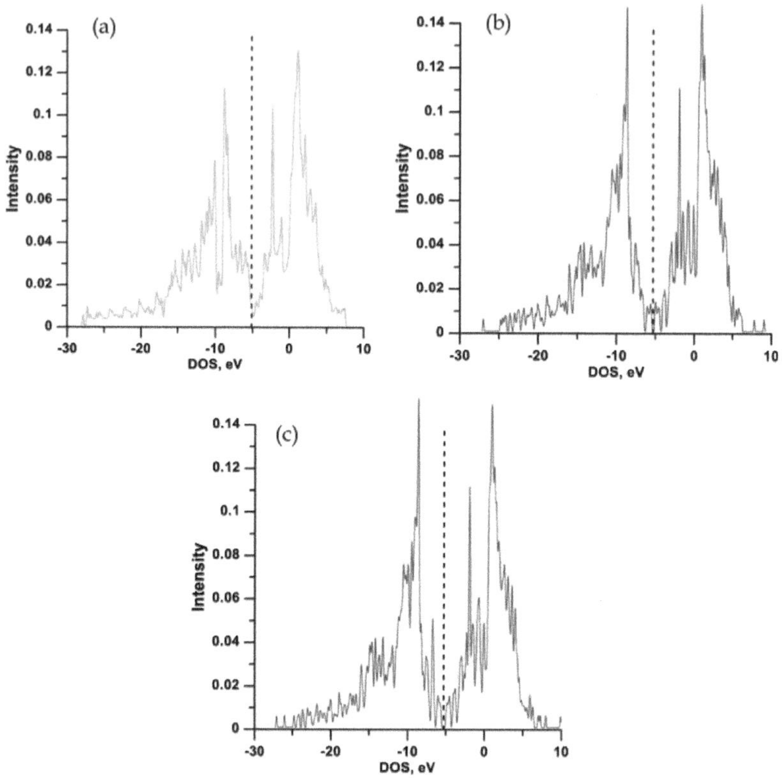

Figure 8.5 DOS spectra of graphene/graphane composites: (a) structure of type AAA; (b) structure of type ABA; (c) structure of type AAB.

that the function has a peak near the Fermi level typical for armchair graphene. Far from the Fermi level, the line becomes non-regular, which can be explained by the contribution of the graphane part. The inset in the graph shows that the transmission function does not touch the x-axis. Thus, the graphene/graphane composite has semiconductive properties.

The transmission functions are shown in Fig. 8.7 (units of measurement $G_0 = e^2/h$). Their graphs qualitatively match the graphs for the composite AAA, but an important difference is observed near the Fermi level. As shown in the highlighted fragment, the transmission function touches the x-axis at the Fermi level. It means that at the Fermi level, the structures of ABA and AAB types

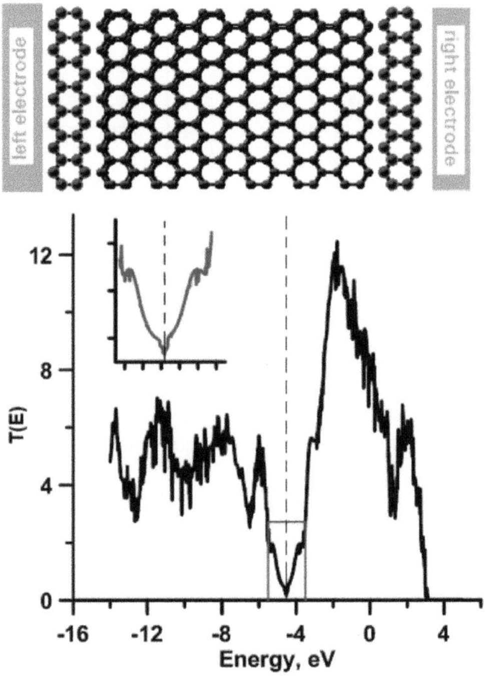

Figure 8.6 Graphene/graphane composites with electrodes as thin graphene nanoribbons (top). Transmission function for the graphene-graphane composite of AAA type. The dotted line stands for the Fermi level (behind).

have no channel of conductivity. The structure of AAA has higher conductivity of linear response.

On the basis of the calculated transmission function $T(E)$, the conductivities of the considered structures were calculated by the following formula:

$$G = \frac{\mathrm{I}}{V} = \frac{e^2}{h} \int_{-\infty}^{\infty} T(E) F_{\mathrm{T}}(E - \mu) dE, \tag{8.1}$$

where e is electron charge, h is Planck's constant, μ is the Fermi level of the structure, and the function $F_{\mathrm{T}}(E)$ can be found by the expression:

$$F_{\mathrm{T}}(E) = \frac{1}{4k_{\mathrm{B}}T} \sec h^2 \frac{E}{2k_{\mathrm{B}}T}, \tag{8.2}$$

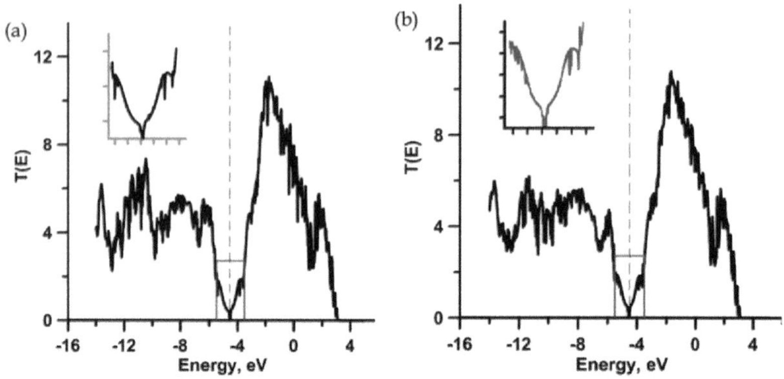

Figure 8.7 Transmission functions for the composites of ABA (a) and AAB type (b). The dotted line stands for the Fermi level.

Table 8.2 Resistance and electrical conductivity of different graphene/graphane composites

Type of graphene/ graphane composite	Resistance (kΩ)	Electrical conductivity (mS)
AAA	9.25	0.108
ABA	7.99	0.125
AAB	7.46	0.131

where k_B is the Boltzmann constant and T is the temperature. The values of resistance and conductivities are showed in Table 8.2.

From the table, it can be seen that the graphene/graphane composite of AAA type demonstrates the lowest resistance, which confirms our suggestion based on the analysis of the transmission function.

8.5 Conclusion

By the method of molecular dynamics on the basis of the AIREBO potential, mathematical models of graphene/graphane composites with different types of packing were derived. The composite of AAA type was the most energy favorable, relying on the values of heat of formation. It was shown that the addition of graphene layers to graphane sharply decreased the energy gap to ~0.5 eV, and the new composite demonstrated semiconductive properties. The calculated values of resistance show that the composites of ABA and AAB types have higher values of electrical conductivity, which make these semiconductive structures prospective materials for application in thin, flexible displays, solar panels, and waveguides.

References

1. Lee, C., Wei, X., Kysar, J. W., and Hone, J. (2008). Measurement of the elastic properties and intrinsic strength of monolayer graphene, *Science*, **321**, pp. 385–388.
2. Chen, Z., Lin, Y., Rooks, M. J., and Avouris, P. (2007). Graphene nano-ribbon electronics, *Phys. E*, **40**, pp. 228–232.
3. Seol, G., Kumar, B., and Guo, J. (2012). Performance projection of graphene nanomesh and nanoroad transistors, *Nano Res.*, **5**, pp. 164–171.
4. Sofo, J. O., Chaudhari, A. S., and Barber, G. D. (2007). Graphane: A two-dimensional hydrocarbon, *Phys. Rev. B*, **75**, pp. 153401.
5. Elias, D. C., Nair, R. R., Mohiuddin, T. M. G., Morozov, S. V., Blake, P., Halsall, M. P., Ferrari, A. C., Boukhvalov, D. W., Katsnelson, M. I., Geim, A. K., and Novoselov, K. S. (2009). Control of graphene's properties by reversible hydrogenation: Evidence for graphane, *Science*, **323**, pp. 610–613.
6. Xiang, H., Kan, E., Wei, S.-H., Whangbo, M. H., and Yang, J. (2009). "Narrow" graphene nanoribbons made easier by partial hydrogenation, *Nano Lett.*, **9(12)**, pp. 4025–4030.
7. Guest, J. R., Bode, M., and Guisinger, N. P. (2009). Patterning graphene at the nanometer scale via hydrogen desorption, *Nano Lett.*, **9(12)**, pp. 4343–4347.
8. Kong, E. S. W. (2015). *Nanomaterials, Polymers and Devices: Materials Functionalization and Device Fabrication* (Australian Institute of Technology, Australia).
9. Antonov, V. E., Bashkin, I. O., Bazhenov, A. V., Bulychev, B. M., Fedotov, V. K., Fursova, T. N., Kolesnikov, A. I., Kulakov, V. I., Lukashev, R. V., Matveev, D. V., Sakharov, M. K., and Shulga, Y. M. (2016). Multilayer graphane synthesized under high hydrogen pressure, *Carbon*, **100**, pp. 465–473.

10. Jones, J. D., Mahajan, K. K., Williams, W. H., Ecton, P. A., Mo, Y., and Perez, J. M. (2010). Formation of graphane and partially hydrogenated graphene by electron irradiation of adsorbates on graphene, *Carbon*, **48**, pp. 2335–2340.
11. Reddy, C. D., Cheng, Q. H., Shenoy, V. B., and Zhang, Y. W. (2011). Interfacial properties and morphologies of graphene–graphane composite sheets, *J. Appl. Phys.*, **109**, p. 054314.
12. Jin, M., Jiao, N., and Da, H. X. (2015). Anisotropic optical properties of graphene/graphane superlattices, *Solid State Sci.*, **40**, pp. 71–76.
13. Averill, F. W. and Morris, J. R. (2011). Graphene/graphane interface energy and implications for defects, *Phys. Rev. B.*, **84**, p. 035411
14. Schmidt, M. J. and Loss, D. (2010). Edge states and enhanced spin-orbit interaction at graphene/graphane interfaces, *Phys. Rev. B*, **81**, p. 165439.
15. Yang, Y. E., Xiao, Y., and Yan, X. H. (2016). Charge distribution of lithium-doped graphane/graphene hybrid system: Role of nearly-free electronic states, *Solid State Comm.*, **229**, pp. 43–48.
16. Zhang, W. X., He, C., Li, T., Gong, S. B., Zhao, L., and Tao, J. Y. (2015). First-principles study on the electronic and magnetic properties of armchair graphane/graphene heterostructure nanoribbons, *Solid State Comm.*, **211**, pp. 23–28.
17. http://nanokvazar.ru/
18. Stuart, S. J., Tutein, A. B., and Harrison, J. A. (2000). A reactive potential for hydrocarbons with intermolecular interactions, *J. Chem. Phys.*, **112**, pp. 6472–6486.
19. Nechaev, Y. S. (2011). The high-density hydrogen carrier intercalation in graphane-like nanostructures, relevance to its on-board storage in fuel-cell-powered vehicles, *Open Fuel Cells J.*, **4**, p. 16.
20. Elstner, M., Porezag, D., Jungnickel, G., Elsner, J., Haugk, M., Frauenheim, Th., Suhai, S., and Seifert, G. (1998). Self-consistent-charge density-functional tight-binding method for simulations of complex materials properties, *Phys. Rev. B*, **58**, pp. 7260–7268.

Chapter 9

New Physical Effects in Graphene/C_{60} Hybrid Systems

Olga E. Glukhova,[a,b] Vladislav V. Shunaev,[a] Vadim V. Mitrofanov,[a] and Michael M. Slepchenkov[a]

[a]*Physical Department, Saratov State University, 83 Astrakhanskaya Street 410010, Saratov, Russia*

[b]*I. M. Sechenov First Moscow State Medical University, Laboratory of Biomedical Nanotechnology, Trubetskaya Street 8-2, Moscow, 119991, Russia*

glukhovaoe@info.sgu.ru

This chapter presents the results of an in silico study of the effect of local hydrogenation of a graphene atomic grid on the character of the displacement of fullerene C_{60} within the graphene substrate. Several topological models of a graphene fragment containing different numbers of hydrogen atoms are considered. In the considered models, the deposited hydrogen atoms were located on one side of the graphene plane and formed various topological patterns. Based on the results of the molecular dynamics experiments, a new method for controlling the shape of the C_{60} trajectory on graphene is proposed by changing the topological variant of the arrangement of hydrogen atoms on graphene. The conditions for the appearance of a circular current on graphene with an ordered motion of a charged C_{60} fullerene are found.

2D and 3D Graphene Nanocomposites: Fundamentals, Design, and Devices
Edited by Olga E. Glukhova

ISBN 978-981-4800-41-9 (Hardcover), 978-0-429-20150-9 (eBook)
www.jennystanford.com

9.1 Introduction

At present, researchers have shown particular interest on the adsorption and desorption of various molecules on graphene [1–6], as well as the optimization and control of molecule/graphene hybrid structures [7–10]. One of the most promising molecular objects used to study the nature of interaction of a molecule with a surface is fullerene C_{60} due to its highly symmetric structure and single-element composition. In turn, the graphene/C_{60} carbon system serves as an ideal model system for studying the nature of interactions at the molecular level, the phenomena of electrotransfer in hybrid structures, as well as for new objects of molecular electronics and spintronics. A number of interesting physical phenomena and effects have been discovered in hybrid graphene/C_{60} molecular systems, which can be used to create new nanodevices. In particular, the phenomenon of charge transfer between C_{60} and graphene has been established [11]. It has been shown that the charge transfer affects the graphene electronic transport and is sensitive to the local orientation of the C_{60} and the presence or absence of an underlying substrate [12]. The ability to fine-tune the charge transfer state of C_{60} interfaced to graphene, for example by electrostatic gating, opens wide prospects for creation on the basis of such heterostructures of new electrochemical and optoelectronic devices. In a number of papers, the phenomenon of the diffusion motion of a C_{60} molecule on a graphene substrate has been investigated [8, 10, 13–15].

The influence of various external factors on the character of C_{60} displacements on graphene, including temperature, external field, substrate corrugation, and chemical functionalization of graphene has been considered. It has been found that under some temperature, fullerene demonstrates two different modes of Brownian motion—ballistic and quasi-continuous—the transition between which is accomplished by varying the temperature [8]. In addition, it has been shown that an increase in the temperature gradient leads to an increase in the speed of the directed movement of C_{60} [14]. At the same time, fullerene starts to lose its mobility on hydrogenated graphene. Moreover, the greater the amount of hydrogen cover on graphene, the lower the mobility of fullerene on it

[15]. The external electric field, on the contrary, forces the fullerene C_{60} to move along the graphene [10], and by varying the magnitude of the field and its direction, it is possible to control the speed of C_{60} and the direction of its displacements [7]. The corrugation of the graphene substrate also makes it possible to control the nature of the displacements of C_{60}. It has been established that the motion of the C_{60} molecule acquires a finite character in the case of giving the substrate a certain corrugation (corrugation wavelength of 3.4 nm and wave depth of 1.6 nm). The application of an external electric field allows us to manipulate the motion of C_{60} along the bottom of graphene [7].

In search of new physical effects and regularities of the behavior of the C_{60} molecule on a graphene substrate, we considered the joint effect of several external factors mentioned above on the dynamics of the C_{60} displacement. This chapter aims to achieve greater controllability of the shape of the trajectory of the fullerene motion. Taking into account the already available data about the ways to affect the C_{60} motion on graphene, we considered a graphene substrate with local hydrogenation of the atomic grid segment. Several topological models of the graphene fragment with various variants of the arrangement of hydrogen atoms on it have been created.

9.2 Computational Details

Numerical experiments were carried out in the Kvazar software package [16] using the molecular dynamics method and the self-consistent charge-density-functional tight-binding method (SCC DFTB) [17]. Within the SCC DFTB method, we considered the function of the reference electron density with some charge fluctuations. A second-order term was added to the expression for the total energy of the system written in terms of density functional theory (DFT). This term takes into account fluctuations in the charge density:

$$E_{\text{SCC-DFTB}} = \sum_{i\mu\nu} c^i_\mu c^i_\nu H^0_{\mu\nu} + \frac{1}{2} \sum_{\alpha\beta} \Delta q_\alpha \Delta q_\beta \gamma_{\alpha\beta} + E_{\text{rep}} \qquad (9.1)$$

where $H^0_{\mu\nu}$ are the Hamiltonian elements; c^i_μ, c^i_ν are the coefficients in the expansion of molecular orbitals as the linear combination of atomic orbitals (LCAO-MO) for the μ and ν-th atomic orbitals in the i-th molecular orbitals; E_{rep} is the energy of the near repulsive interaction; $\gamma_{\alpha\beta}$ is the analytic function describing the effective local electron–electron interaction; Δq_α and Δq_β are the fluctuations of the atomic Mulliken charges; the indices α and β denote atoms on which the wave functions are centered. The trajectory of atoms was described by the molecular dynamics method [18]. Verlet algorithm was used for numerical integration. The step of numerical integration was 1 fs. To maintain a constant temperature in the course of numerical experiments, the Berendsen thermostat was used [19].

9.3 Behavior of Fullerene C_{60} on Graphene with Local Hydrogenation of the Atomic Grid Segment

9.3.1 Graphene Fragment with a Single Hydrogen Atom

The object of investigation was a graphene sheet of length 30 Å along the zigzag direction and 33 Å along the armchair direction. At first we considered the chemisorption of a single hydrogen atom on a 5-7-7-5 defect. This defect led to the deformation of graphene, as shown in Fig. 9.1a. The interaction energy profile for the C_{60} molecule with the graphene sheet is shown in Fig. 9.1b. It can be seen that the energy peak appeared in the area of the local defect.

Further, we considered the dynamical behavior of the C_{60} molecule on graphene of a given structure topology. To identify the regularities of the behavior of fullerene, we carried out a series of numerical experiments using molecular dynamics at a temperature in the range of 50–300 K. Calculation results showed that the behavior of fullerene had a complicated character in all considered cases. The C_{60} molecule moved from the upper edge of graphene sheet along the y-axis with a small deviation of straight-line trajectory. Achieving the area of localized defect, the object deviated from the initial trajectory and rounded the defect. The

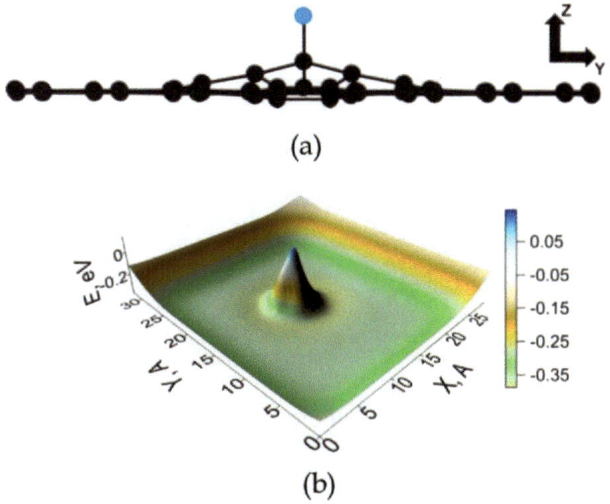

Figure 9.1 The influence of local defect in the form of the chemisorption of a single hydrogen atom on a graphene–C_{60} system: (a) on the atomic structure; (b) on the interaction energy profile for fullerene with graphene.

reason for such a behavior of the molecule is the repulsive barrier that occurs during the process of interaction between fullerene and hydrogen atom. After the passing of defect, the fullerene gradually returns to the initial track. Figure 9.2 illustrates the typical case of behavior of fullerene on graphene of the considered topology at 290 K. Thus, at this stage of investigation, it was concluded that the creation of localized defect in the form of the chemisorption of a single hydrogen atom on a 5-7-7-5 defect in the graphene atomic grid only for a while allowed us to deflect the fullerene from the initial course while it goes around the obstacle in the way. So we need another modification of graphene's atomic structure to give a circular character to the motion of fullerene.

9.4 Graphene Fragment with a Local Segment Containing 19 Hydrogen Atoms

In order to search the optimal structural modification of the graphene sheet that provides the desired effect, we considered the

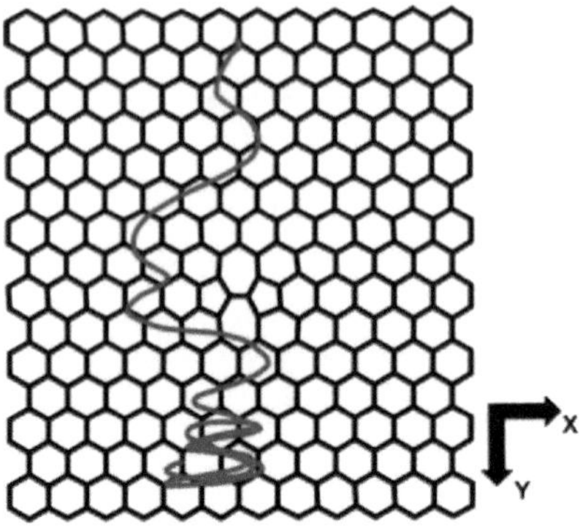

Figure 9.2 The trajectory of the center of mass of C_{60} at 290 K.

case of local graphene hydrogenation by a group of 19 atoms. These atoms were chemically absorbed in the central area of the graphene atomic grid. The creation of such a defect deformed the graphene sheet. The graphene sag in the area of hydrogenation is shown in Fig. 9.3a. Figure 9.3b shows the interaction energy profile for the C_{60} molecule with the graphene sheet for this case. It can be seen that there is a convex area in the center of the graphene sheet. The sizes of this area match the sizes of the hydrogenation area.

In order to identify regularities in the dynamical behavior of the C_{60} molecule on a graphene platform, we made a series of numerical experiments at temperatures in the range of 50 to 300 K. In the course of an investigation, it was found that achieving the area of hydrogenation defect, the fullerene started to move just within this structure site due to the repulsive energy barrier. Moreover, it was found that at a definite temperature, the motion of fullerene within the aforementioned site becomes cyclic. According to our calculations, this effect appears at 160 K. Figure 9.4 shows the trajectory of the center of mass of C_{60} in this case.

It can be seen that the motion of fullerene had an irregular character, which was accompanied by multiple attempts to describe a loop form of trajectory. With increasing temperature, C_{60} began to chaotically move inside the graphene bowl without loop trajectory. Thus, in order to give more directional character to the loop motion observed at 160 K and to increase the repeatability of such a

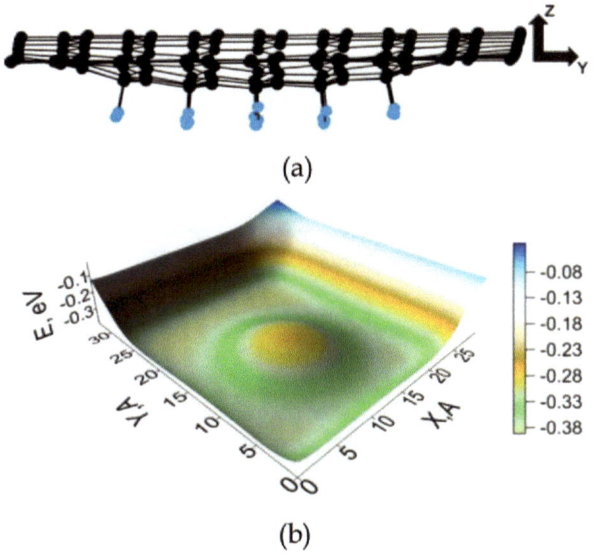

Figure 9.3 The influence of local defect in the form of chemisorbed hydrogen atoms on a graphene–C_{60} system: (a) on the atom structure; (b) on the interaction energy profile for C_{60} with graphene.

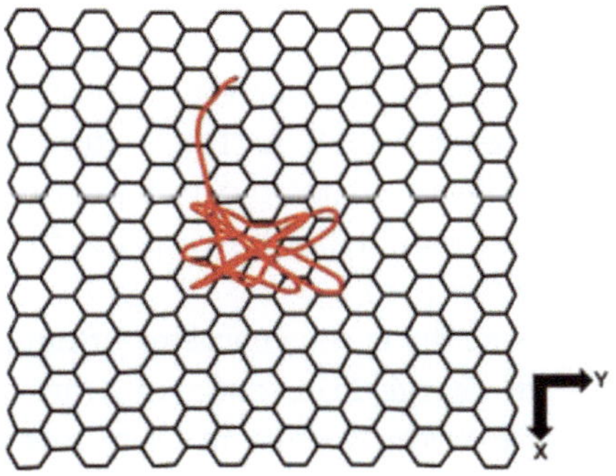

Figure 9.4 The trajectory of the center of mass of C_{60} under 160 K.

trajectory, we need to add another factor. We used an external electric field with a strength of 1×10^6 V/m in order to give a cyclic character to the motion of fullerene. Field lines were directed along the y-axis of the graphene plane. The fullerene was preliminary charged to $+1$e by ion encapsulation. The analysis of the trajectory

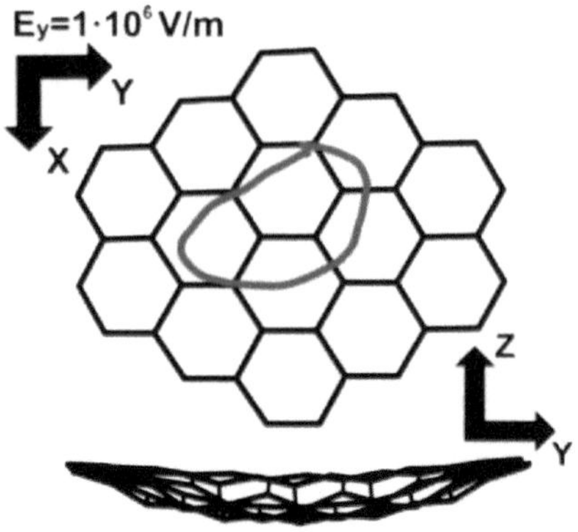

Figure 9.5 The highlighted fragment of the trajectory of motion in the loop form for fullerene in the bottom of a graphene bowl with the diameter $d = 9$ Å.

of C_{60} showed that within the localized defect area, the motion of fullerene acquires a regulated character due to the electric field. Since a certain point of time (~80 ps), C_{60} makes a cyclic motion with some periodicity. The important role of the electric field should be noted. By setting the electric field vector along the *y*-axis, we forced fullerene to make a clockwise cyclic rotation.

Figure 9.5 shows the highlighted fragment of the trajectory of motion in the loop form for fullerene in the bottom of a graphene bowl with the diameter $d = 9$ Å. The picture of loop with an unideal circle can be caused by the topological features of the graphene atomic grid within the bowl, primarily by its strongly marked curvature in this area.

Taking into account the cyclic character of motion for fullerene and the presence of charge in it, one can draw a direct analogy with the motion of electron within an atom in circle orbits [20]. Such a motion is equivalent to the closed current loop. Thus, it can be assumed that the phenomena of a current flow can occur during the cyclic motion of C_{60} within the area of hydrogenation defect localization. We gave a numerical evaluation of cyclic current induced in the graphene–C_{60} system.

The fullerene moved on a closed trajectory, which was represented by an ellipse with the length of semi-major axis $a = 3.75$ Å and the length of semi-minor axis $b = 2$ Å. The length of the ellipse was equal to 18.48 Å, and the area of the ellipse was 23.56 Å^2. The current was calculated according to the known formula $I = e\upsilon = ev/2\pi r$, where e is the electron charge, υ is the number of turns per second, v is the average velocity of fullerene mass center motion, and r is the radius of the circular orbit. Since the average velocity of the C_{60} molecule was ~65 m/s in the considered case, the circular current can be estimated as ~6 nA according to the aforementioned formula.

We also calculated the magnitude of the magnetic field B created by the closed current loop. This loop represented the closed elliptical trajectory for the fullerene with charge +1e. The maximum magnitude of the magnetic field B was achieved in the center of the ellipse. The magnetic field decreases when moving away from the current loop, as shown in Fig. 9.6. Thus, $B = 0.822$ mkT at the point O. The field decreases when moving along the straight line OA. The magnetic field becomes zero at the distance of ~0.6 nm away from the ellipse. The magnetic field B was calculated according to the well-known formula for a circle current where the focal parameter of an ellipse was used instead of the radius of a circular ring. Figure 9.6 shows this formula with the account of geometric parameters of the aforementioned ellipse and the current magnitude. Taking into account the area of the current loop, we can assume the magnitude of the magnetic flux. For example, the maximum magnitude of the magnetic flux was equal to 0.7 nWb for a graphene surface of 1 mm^2.

9.5 Graphene Fragment with a Local Segment Containing 44 Hydrogen Atoms

After the detection of the conditions for the occurrence of the circular current in the graphene–C_{60} system, it was necessary to find a way to exceed its value. The performed numerical simulations allowed us to determine that the increase in the external field strength forces the fullerene to leave the circular orbit and to stop the current flow. So we got the idea to decide the noted problem by varying the geometrical sizes of the graphene bowl by using a bigger amount of chemically adsorbed hydrogen atoms on graphene.

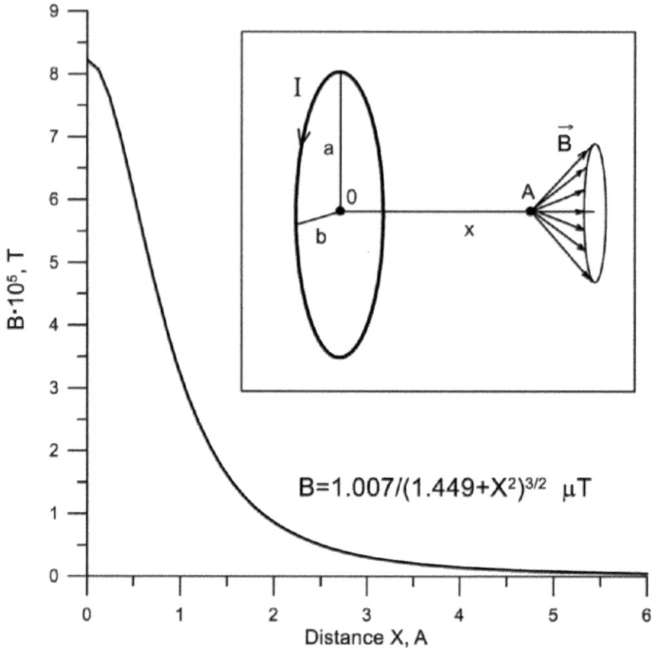

Figure 9.6 The change in the magnetic field B generated by the closed current loop as the distance away from the center of the loop.

To check this idea, we conducted a series of numerical experiments for the local defect of the chemisorption of 44 hydrogen atoms on graphene. This defect created a graphene bowl with a bigger diameter $d = 14.6$ Å. Figure 9.7 shows the interaction energy profile for the fullerene C_{60} with graphene in this case. It is seen that the energy relief has undergone significant changes. In the center of the profile, there is the energy well with a depth of 1.8 eV unlike the previous case for 19 hydrogen atoms, where the convex region was observed. Thus, at a definite concentration of chemically absorbed hydrogen atoms on graphene, the convex energy profile transforms to the profile with the energy well.

As described in the previous case for the bowl of lower diameter, at the first stage of further investigation, we revealed the temperature that gives a loop form of trajectory. Such a temperature was equal to ~175 K. Further, we applied an external electric field

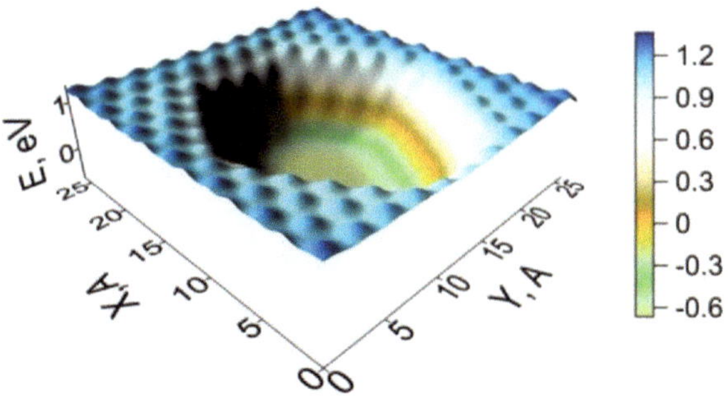

Figure 9.7 The interaction energy profile for the fullerene with graphene, which contains the localized defect of the chemisorption of 44 hydrogen atoms.

of strength 1×10^6 V/m along the *y*-axis of the graphene plane. After the application of the external electric field, the trajectory of the motion of fullerene makes a loop and repeats after about 20 ps. Figure 9.8 shows the fragments of the C_{60} loop trajectories in two consecutive time intervals for the given case.

The figure shows that in the course of time, the fullerene revolves in elliptical trajectories in the center of the graphene bowl. Herewith, the loop trajectory of the fullerene rotates clockwise according to the direction of the applied field. The circle current was calculated for one of the aforementioned closed trajectories that represented the ellipse with the length of semi-major axis $a = 3.32$ Å and the length of semi-minor axis $b = 1.28$ Å. The area of the ellipse was 13.35 Å^2. The calculated circle current was ~10.4 nA. For this case, we calculated the magnetic field *B* generated by the closed loop in the form of an ellipse. Figure 9.9 shows the change in the magnetic field *B* as the distance away from the center of the loop.

It can be seen that the magnetic field decreases and achieves a maximum magnitude in the center of an ellipse ($B = 1.2$ mkT). So the increase in the diameter of the graphene bowl just by 5.6 Å allowed us to obtain almost a twofold increase in the magnitude of the circular current in the closed loop. The magnetic field produced by the circular current loop has also increased. Thus, the suggested way of enhancing the observed physical effect is effective and can be realized in practice. It is predicted that by choosing appropriate

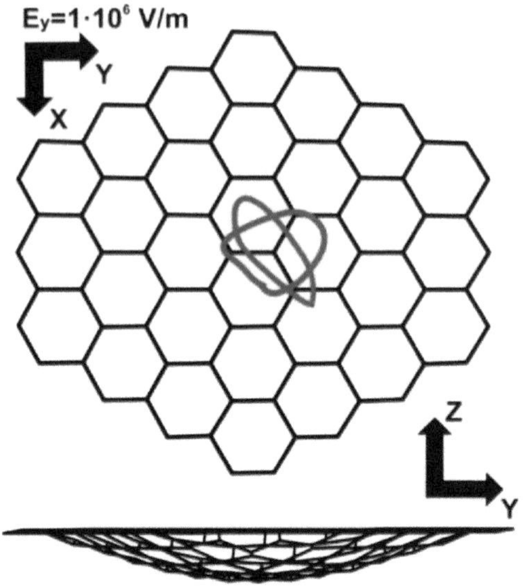

Figure 9.8 The fragments of the C_{60} loop trajectory in the bottom of the graphene bowl with the diameter d = 14.6 Å in two consecutive time intervals.

sizes of graphene bowl, it is possible to increase the magnitude of current by a factor of tens and hundreds.

9.6 Controlling the Shape of the C_{60} Trajectory by Varying the Topological Pattern of the Arrangement of Hydrogen Atoms on Graphene

The next stage of our study was the search for optimal variants of the arrangement of hydrogen atoms in the graphene plane, at which C_{60} could move along a trajectory assuming a strictly prescribed shape. We considered two topological variants of graphene with deposited hydrogen atoms. These topological variants differ by an arrangement of hydrogen atoms on graphene. In the first case, hydrogen atoms were arranged in two rows along the graphene sheet, forming a corridor (Fig. 9.10). In the second case, the

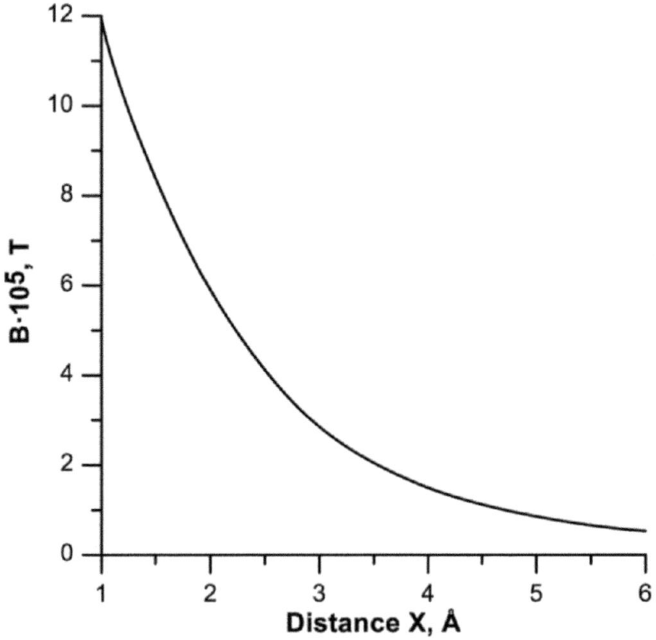

Figure 9.9 The change in the magnetic field B generated by the closed current loop as the distance away from the center of the loop.

deposited hydrogen atoms were arranged in a circle (Fig. 9.11). We will refer to the above-mentioned structures as hybrid structures of the first (I) and second (II) types, respectively.

The energy profile of the interaction of graphene and fullerene was calculated for each of the considered graphene/C_{60} hybrid systems. Figure 9.12 shows the energy profile for the graphene/C_{60} hybrid structure of type I. It can be seen from this figure that the energy profile has pronounced peaks at the sites of graphene atomic network, where hydrogen atoms deposited. These peaks indicate energy barriers for fullerene C_{60} moving along graphene. The energy profile of the interaction of graphene and fullerene for the structure of type II is shown in Fig. 9.13. As in the previous case, at the sites of hydrogen atoms, the profile has pronounced peaks, which are an energy barrier for fullerene atoms.

According to our hypothesis, to change the C_{60} movement trajectory, one can use the first topological variant for structures of

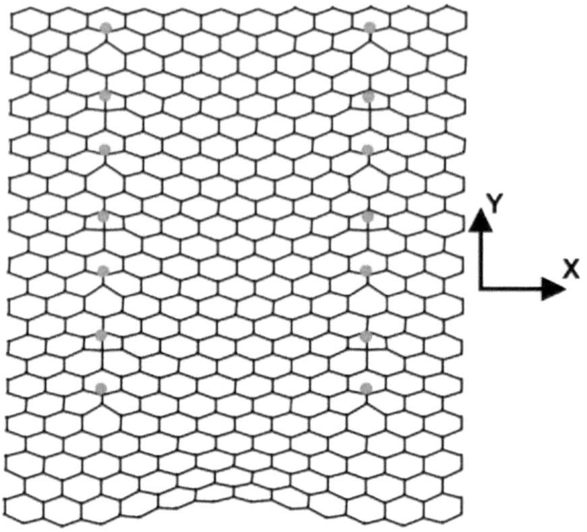

Figure 9.10 Atomic structure of the investigated graphene/C_{60} hybrid system of type I.

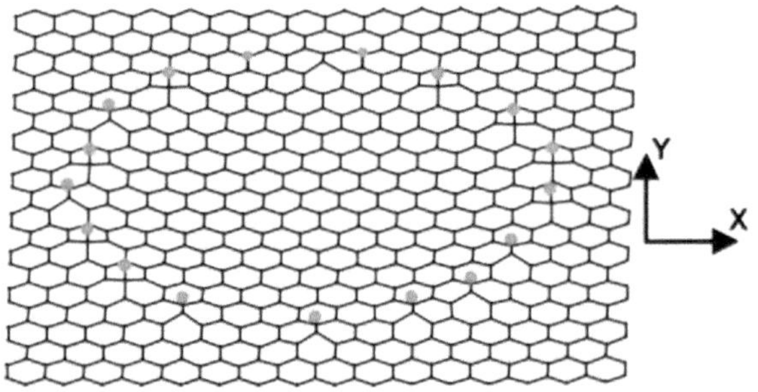

Figure 9.11 Atomic structure of the investigated graphene/C_{60} hybrid system of type II.

the first type; the goal was to obtain the trajectory of the motion of the fullerene object along the straight line; for structures of the second type, the goal was to obtain the circular trajectory.

Next, we investigated the dynamics of the C_{60} molecule in the hybrid structures of type I. At the initial time, C_{60} was at the edge of the graphene sheet. At first, the behavior of fullerene was considered only under the influence of temperature, which varied in the range of

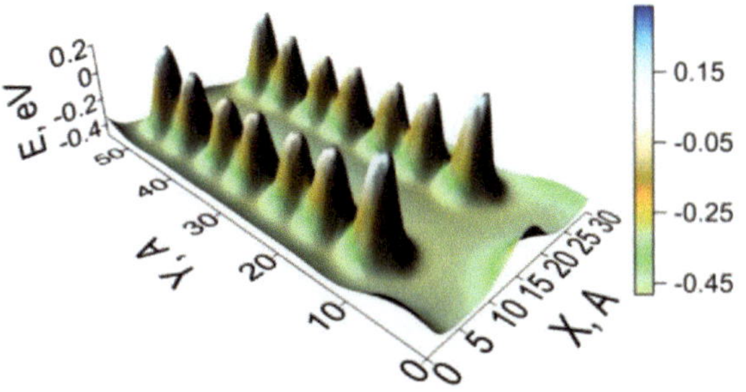

Figure 9.12 Energy profile of the interaction of graphene and fullerene for hybrid structure of type I.

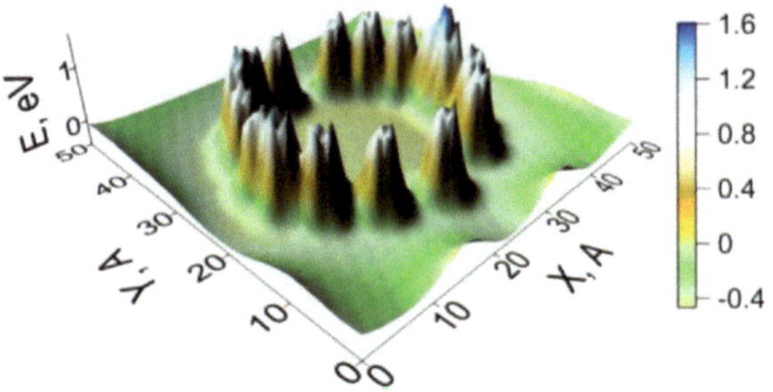

Figure 9.13 Energy profile of the interaction of graphene and fullerene for hybrid structure of type II.

100–300 K. In the course of the investigation, it was established that under the action of a single temperature, fullerene does not move between two identical rows of hydrogen atoms but bends them on the left or right side on account of the presence of a repulsive barrier. One of these situations is shown in Fig. 9.14.

Consequently, it is necessary to give a fullerene impulse, under the action of which it would enter the space between the rows of hydrogen and begin to move in a given direction. As was shown earlier in other scientific papers, such an effect on fullerene can be exerted by means of an external electric field. Thus, our next task was to select the magnitude of the external field strength, at which the object would move inside the gauge created by us. As a result of numerical experiments, we found that this value is 1×10^7 V/m. The motion trajectory for this case is shown in Fig. 9.15.

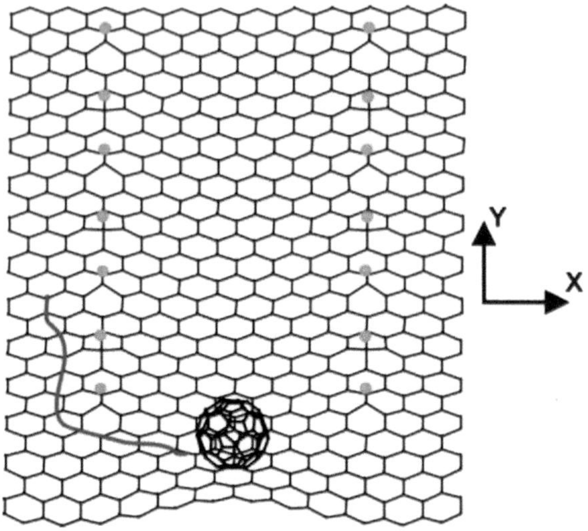

Figure 9.14 Trajectory of the C_{60} motion on graphene in structures of type I under the influence of temperature $T = 300$ K.

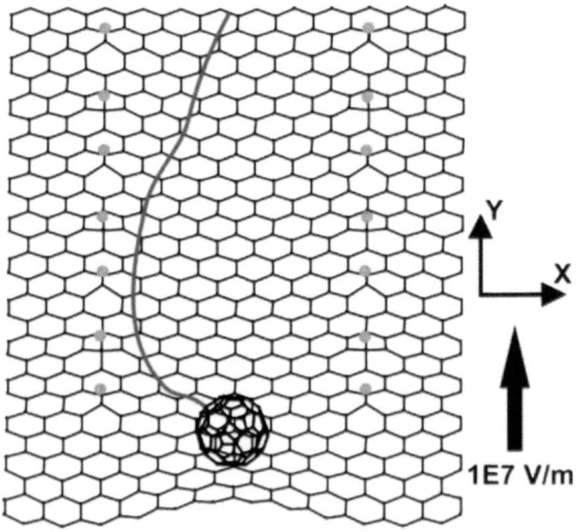

Figure 9.15 Trajectory of the motion of C_{60} on graphene in structures of type I under the influence of a temperature of 300 K and an external electric field with a strength of 1×10^7 V/m.

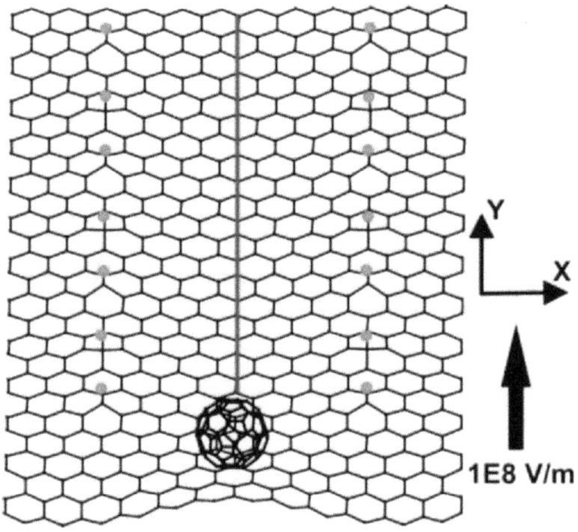

Figure 9.16 Trajectory of the C_{60} motion on graphene in structures of type I under the influence of 300 K temperature and an external electric field with a strength of 1×10^8 V/m.

Analyzing the trajectory, one can see that the fullerene moves between the rows of hydrogen atoms with some deviation from the straight line. To reduce the magnitude of the deflection, the field strength increased up to 1×10^8 V/m. The motion trajectory for this case is shown in Fig. 9.16. It is clearly seen from the figure that fullerene moves between rows of hydrogen in a straight line.

Further numerical experiments were carried out to reveal the character of the dynamics of C_{60} in the hybrid structures of type II. Under the action of only the temperature ($T = 300$ K), the fullerene again departed from the rut created for it. This is clearly seen from the trajectory shown in Fig. 9.17.

However, an external field of intensity 1×10^7 V/m directed the fullerene inside a graphene section containing hydrogen atoms, where C_{60} began to move along the circular trajectory (Fig. 9.18).

9.7 Conclusion

A new effective way to manipulate the fullerene molecules was proposed based on the results of numerical modeling of the behavior

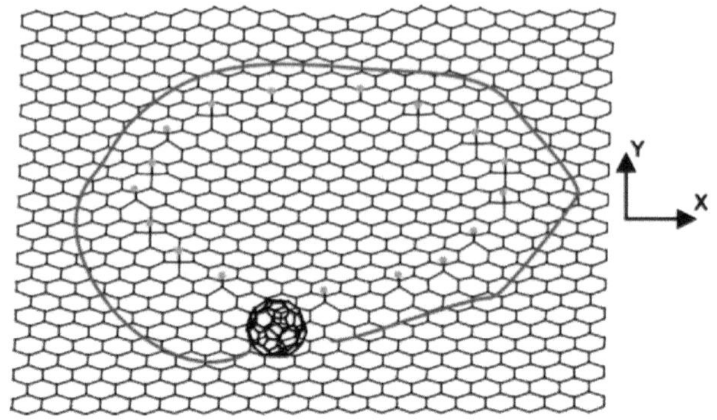

Figure 9.17 Trajectory of the C_{60} motion on graphene in structures of type II under the influence of temperature equal to 300 K.

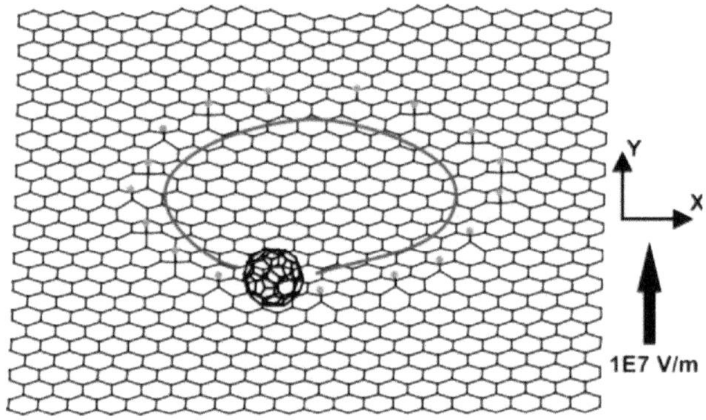

Figure 9.18 Trajectory of the C_{60} motion on graphene in structures of type II under the influence of 300 K and an external electric field of 1×10^7 V/m.

of fullerene C_{60} on a graphene substrate with the defects of the local hydrogenation. It was shown that the chemisorption of the group of hydrogen atoms on a graphene sheet created a bowl in the center of the graphene atomic grid. The finite motion of the fullerene was observed inside this bowl. The minimal depth of the bowl's energy profile should be equal to 0.3 eV for the finite motion of the fullerene object. It was established that at the determined temperature and strength of the external electric field, the fullerene starts to move inside the bowl in elliptic trajectories. In the course of the investigation of the C_{60} circular motion, we found a new physical effect of the circular current loop. The magnitude of this current was equal to ~10.4 nA at a temperature of 175 K and an electric field strength of 1×10^6 V/m. As an effective way of enhancing current, we proposed to vary the sizes of the graphene bowl. It can be assumed that the models of new miniature power sources for a variety of nanodevices will be developed on the basis of the obtained results.

A new method for controlling the shape of the C_{60} trajectory on graphene was proposed by varying the topological drawings of the arrangement of hydrogen atoms on graphene. It was established that by arranging the hydrogen atoms on graphene in a certain way, one can achieve the motion of fullerene C_{60} in the localized region of the graphene sheet due to the energy barrier between carbon and hydrogen atoms. For the investigated graphene/C_{60} hybrid system of type I, the mode of displacement of fullerene C_{60} on graphene along the straight line was achieved at a temperature of 300 K and the action of an external field of intensity 1×10^8 V/m. For the graphene/C_{60} hybrid system of type II, the mode of displacement of fullerene C_{60} on graphene along a circular trajectory was achieved at a temperature of 300 K and an external field strength of 1×10^7 V/m. This method allowed us to consider the motion of fullerene C_{60} only within the localized graphene region, and the electric field served as an external driving factor, due to which the fullerene rises to a given site.

References

1. Jeon, E. K., Yang, C. S., Shen, Y. F., Nakanishi, T., Jeong, D. S., Kim, J. J., Ahn, K. S., Kong, K. J., and Lee, J. O. (2012). Photoconductivity and enhanced memory effects in hybrid C_{60}-graphene transistors, *Nanotechnology*, **23**, pp. 455202.
2. Wang, R., Wang, S., Wang, X., Meyer, J. A. S., Hedegård, P., Laursen, B. W., Cheng, Z., and Qiu, X. (2013). Charge transfer and current fluctuations in single layer graphene transistors modified by self-assembled C_{60} adlayers, *Small*, **9**, pp. 2420.
3. Yi, Y., Choi, W. M., Son, B., Kim, J. W., and Kang, S. J. (2011). Interfacial electronic structures between fullerene and multilayer graphene for N-type organic semiconducting devices, *Carbon*, **49**, pp. 4936.

Chapter 10

Graphene/Carbon Nanotube Hybrid Material as an Element Base of Optical Nanodevices

Igor S. Nefedov[a] and Olga E. Glukhova[b,c]

[a]*Laboratory Nanooptomechanics, ITMO University, Lomonosova Street 9, 197101, St. Petersburg, Russia*
School of Electrical Engineering, Aalto University, PO Box 13000, 00076 Aalto, Finland
[b]*Physical Department, Saratov State University, 83 Astrakhanskaya Street 410010, Saratov, Russia*
[c]*I. M. Sechenov First Moscow State Medical University, Laboratory of Biomedical Nanotechnology, Trubetskaya Street 8-2, Moscow, 119991, Russia*
glukhovaoe@info.sgu.ru

This chapter studies the electrodynamic properties of a new graphene/carbon nanotube composite material. Topological models of mono- and bilayer composites with nanotubes of metallic and semiconductor conductivity and different intertube distances have been considered. The spectra of the real and imaginary parts of the diagonal elements of the surface conductivity tensor have been obtained. The absorption coefficients for p-polarized and s-polarized radiation have been calculated.

2D and 3D Graphene Nanocomposites: Fundamentals, Design, and Devices
Edited by Olga E. Glukhova

ISBN 978-981-4800-41-9 (Hardcover), 978-0-429-20150-9 (eBook)
www.jennystanford.com

10.1 Introduction

New nanomaterials based on graphene and carbon nanotubes (CNTs) are extremely promising in optical communication. In the recent years, active research has been carried out demonstrating the unique properties of these composites and their great prospects as electro-optical materials for optical modulators of various types [1–4]. In particular, one of these composites is a 2D hybrid composite based on graphene monolayers and nanotubes covalently bonded to them [5–8]. Experiments have proved that a hybrid 2D film demonstrates high efficiency as a photosensitive element of a photodetector in the range of 100–700 nm. It has been already established that one photon absorbed by the film induces a flux of 10^5 electrons. In this case, the response time of the photosensitive element is ~100 μs [9]. The seamless contact between CNTs and the graphene monolayer [10, 11] facilitates the transport of electrons without scattering on defects.

This chapter presents the results of modeling of properties and exploring the prospects of the new hybrid CNT/graphene 2D composite as an element base for new-generation optical modulators. The basis of the in silico research is the popular and reliable quantum-mechanical SCC DFTB 2 method [12, 13].

10.2 Computational Details

The CNT/graphene composite film was modeled as two graphene monolayers and nanotubes between them. We took nanotubes of the same diameter and located them at the same distance from each other. The tube type was zigzag since it was previously shown that composites with tubes of the type (n, 0) zigzag (n = 10, 12, 14, 16, 18, 20) were thermodynamically stable at the tube step of 9–15 hexagons (with step equal to one) [14]. To study the properties of the composite film, we took two groups of CNT/graphene composite models: (1) models with metallic SWCNTs (12,0) and (18,0) and (2) models with semiconducting SWCNTs (14,0) and (16,0). These tubes were not chosen by chance. Earlier, they were determined by experimental studies [11] during the synthesis of the CNT–graphene composite.

First of all thermodynamically stable supercells of the composite were built by the original "Method of magnifying glass" [14]. The supercell was constructed in the following steps: (1) the initial configuration of the CNT/graphene composite was built, including at least seven long nanotubes covalently bonded to the graphene monolayer; (2) the atomic structure was empirically optimized by the AIREBO method; (3) the inner part was cut out and re-optimized by the quantum method; (4) the supercell was cut out, and its translation vectors were determined and again re-optimized by the quantum method. Figure 10.1 shows the obtained supercell (a) and the composite's fragment (b) formed by the nanotubes (18,0) with the step of 13 hexagons. The supercell is translated along the *x* and *y* directions. The thickness of the film (along the *z*-axis) is determined by the tube diameter and the degree of strain. The tubes were slightly deformed, and the graphene became corrugated. The tubes were covalently bonded to the monolayer of the graphene, as shown in Fig. 10.1. Pink balls mark the CNT atoms, and green balls mark graphene. For all the composite models, the length of the C–C bond between CNTs and graphene was ~0.16 nm [14]. Figure 10.1c shows the supercells of mono- and bilayer composite on the base of tubes (10,0) with the tube step of 12 hexagons. It is clearly visible that the middle monolayer of graphene was straightened, bonding the tubes with each other.

To model the interaction of the composite with the incident electromagnetic wave of the IR–visible–UV range, we employed the theory of the linear response of the electron population to the incident electromagnetic wave and Maxwell's equations. Figure 10.2 shows the variant of the configuration for the incident wave vector plane with respect to the supercell of the composite. In the figure, the wave vector **k** of the plane electromagnetic wave is marked with black. The wave vector lies in the *yz* plane. The angle θ corresponds to the angle of incidence, and the vectors **E** and **H** correspond to the electric and magnetic field strengths, respectively. The electrodynamic medium is vacuum. The vector **H** lies in the *yz* plane, and the vector **E** is perpendicular to the plane of incidence wave and is parallel to the CNT axis. In this configuration, the wave

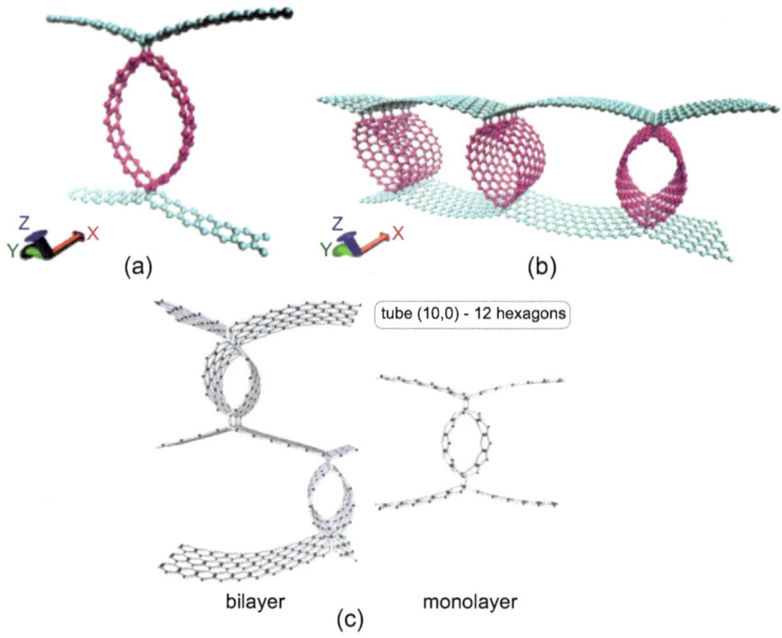

Figure 10.1 Topological model of the graphene/CNT 2D composite with CNT (18,0): (a) supercell; (b) a fragment of the composite (pink color marks CNT atoms, and green marks graphene); (c) supercell.

is a TE wave. In the other case, when the vector **E** is in the YZ plane and the vector **H** is directed along the x-axis (along the axis of the nanotube), the incident wave is a TH wave.

To determine the reflection, transmission, and absorption parameters, Maxwell's equations for electric and magnetic fields in vacuum were solved. Herewith, the thin composite CNT/graphene film is the interface. In this case, it was assumed that the electromagnetic wave passed from the vacuum into the film (half-space $z > 0$) and returned to the vacuum (half-space $z < 0$). To find the field at the interface, boundary conditions were set:

$$\vec{n} \times (\vec{E}_1 - \vec{E}_2) = 0, \vec{n} \times (\vec{H}_1 - \vec{H}_2) = \vec{j} \qquad (10.1)$$

where $\vec{j} = \sigma_{xx} E_x + \sigma_{yy} E_y$ is the surface current density in the composite induced by the incident radiation. The subscripts 1 and 2 refer to the fields in the half-spaces $z > 0$ and $z < 0$, respectively;

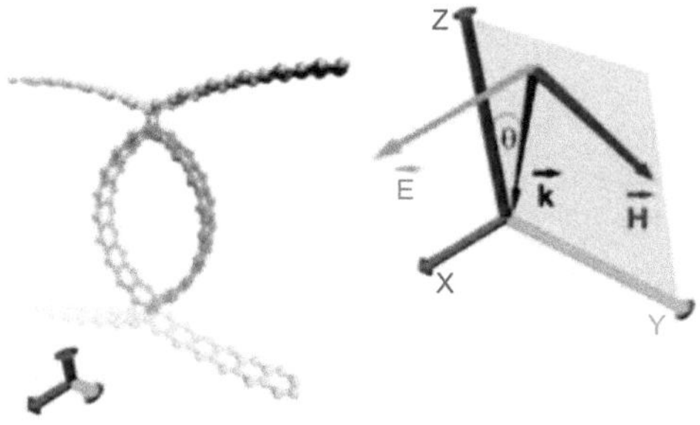

Figure 10.2 Supercell of a graphene/CNT 2D composite and an incident TE wave.

n is the normal vector to the surface. Further, the boundary problem for TE and TH waves will be solved separately.

First of all the elements of the dynamic conductivity tensor were calculated. The complex dynamic electrical conductivity was calculated by the Kubo–Greenwood formula [15], which determined the conductivity as a function of the photon energy Ω and which can be written in the form [16]:

$$\sigma_{\alpha\beta} = \frac{2e\hbar}{im_e^2 S_{\text{cell}}} \frac{1}{N_{\vec{k}}} \times \sum_{k\in BZ}^{N_k} \sum_{m,n} \frac{\widehat{P}_{\alpha}^{nm}(\vec{k}) \cdot \widehat{P}_{\beta}^{mn}(\vec{k})}{E_n(\vec{k}) - E_m(\vec{k}) + \Omega + i\eta} \times \frac{f_\beta[E_n(\vec{k}) - \mu] - f_\beta[E_m(\vec{k}) - \mu]}{E_n(\vec{k}) - E_m(\vec{k})} \tag{10.2}$$

where $f_\beta = 1/(1 + \exp[\beta(x - \mu)]$ is the Fermi–Dirac function with chemical potential u; β is the value reverse to heat energy $\beta = 1/k_{\text{B}}T$; S_{cell} is the area of a unit cell; $N_{\vec{k}}$ is the number of k-points needed for bypassing the Brilloune zone; $\hat{P}_{\alpha}^{nm}(\vec{k})$ are matrix elements corresponding to the α-components of the impulse vector operator; $E_{n/m}(\vec{k})$ is the energy of the valence/conductive zone; m_e and e are the mass and charge of the free electron. The factor 2 takes into account the spin degeneracy. The phenomenological parameter η characterizes electron scattering. To calculate the elements of the pulse matrix, a known relation $\hat{P}(\vec{k}) \to (m_e/\hbar)\nabla_{\vec{k}}\hat{H}(\vec{\vec{k}})$ was used where $\hat{H}(\vec{k})$ is the Hamiltonian. Calculation of the momentum operator matrix elements is described in Ref. [17]. The Hamiltonian was built by the SCC DFTB 2 method.

10.3 Results and Discussion

Figure 10.3 shows the spectra of the real and imaginary parts of the diagonal elements of the dynamic conductivity tensor (which is often called optical due to the range) for four topological models of the composite with the tube step of 13 hexagons and four types of CNTs. Figures 10.3a,c show the graphs for the direction perpendicular to the axis of the tube (*yy*), and Figs. 10.3b,d show the graphs for the direction of the tube's axis (*xx*). The analysis of the spectral profile for both tensor elements shows the presence of released peaks at wavelengths of 190–260 nm. The appearance

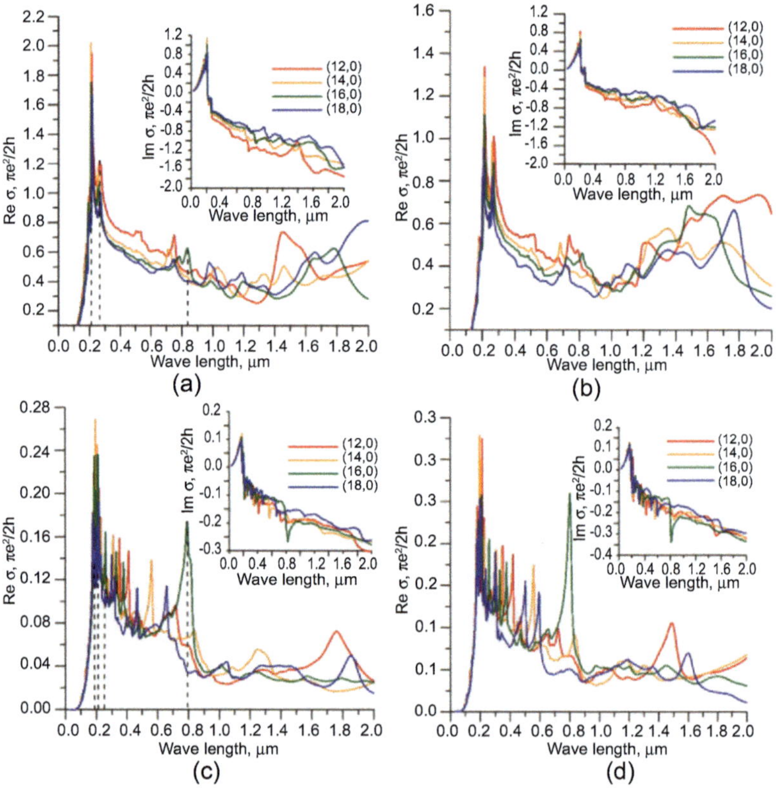

Figure 10.3 Optical conductivity of CNT/graphene composite for the tube step of 13 hexagons for TE (a) and TH waves (b) and for tube step of 9 hexagons for TE (c) and TH waves (d).

of these peaks is caused by the addition of pure graphene in the composite, so these peaks have a high intensity for all the topological models of the studied composite. At the same time, the spectral profile of σ_{yy} is similar to the spectrum of graphene, while the spectral profile of σ_{xx} has a complex character and a set of peaks. As it was shown earlier in Ref. [14], the complex profile of the conductive spectrum along the CNT axis is explained by the influence of nanotubes. The appeared numerous peaks are typical for the conductivity spectrum of isolated individual CNTs. We also note that the intensity of the maximum peak observed at the frequency of 6 eV (206.6 nm) in pure free graphene decreases with the appearance of graphene's corrugation during the formation of the composite. The intensity of peaks for a CNT/graphene film is higher than for pure graphene and isolated nanotubes (for more details, see Fig. 6 in Ref. [14]).

The peak of high intensity in Fig. 10.3b for the model with a tube (16,0) at the wavelength of 800–830 nm attracts attention. It can be predicted that at this wavelength, this composite expects unusual properties during interaction with the incident electromagnetic wave, in particular, during the reflection and absorption of the wave. Let us note that for other cases with other tube steps, the pattern of the spectrum topography remains the same with minor changes.

To compare the properties of mono- and bilayer composites, Fig. 10.4 shows the spectra of optical conductivity. The qualitative similarity of the conductivity spectra can be seen. This is explained by the presence of the same tubes in the composites. They determine the position of the peaks. A clear peak is observed for the real part of the conductivity at a wavelength of 1.5 nm for models with the tube step of 10 hexagons. This peak is clearly caused not by the tubes but by the certain step between them. It is also characteristic that the peak appeared in the region where graphene has no maxima. That is, in contrast to graphene, the composite has an enhanced absorption capacity in the IR region. The composite is capable of absorbing waves in the extended range in comparison to individual tubes and ideal graphene. We also studied the graphs of the complex dielectric constant. Its value, as it is well known, can be easily calculated from the conductivity value $\varepsilon(\omega) = i\sigma(\omega)/\varepsilon_0\omega$; therefore, we give the

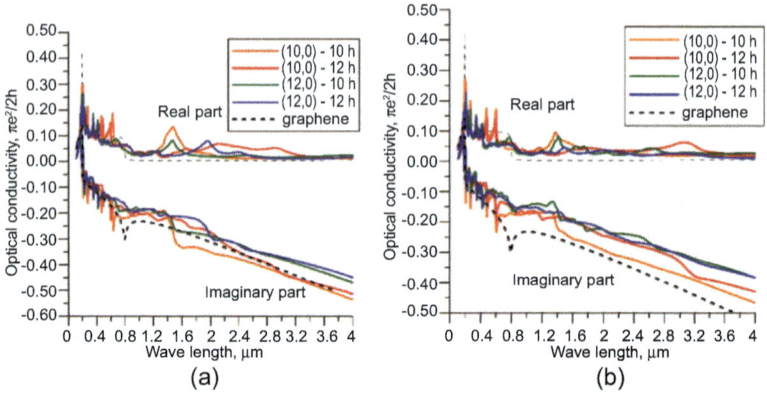

Figure 10.4 Complex optical conductivity of the CNT/graphene composite for mono- (a) and bilayer (b) graphene of various topological models (the element of the conductivity tensor XX).

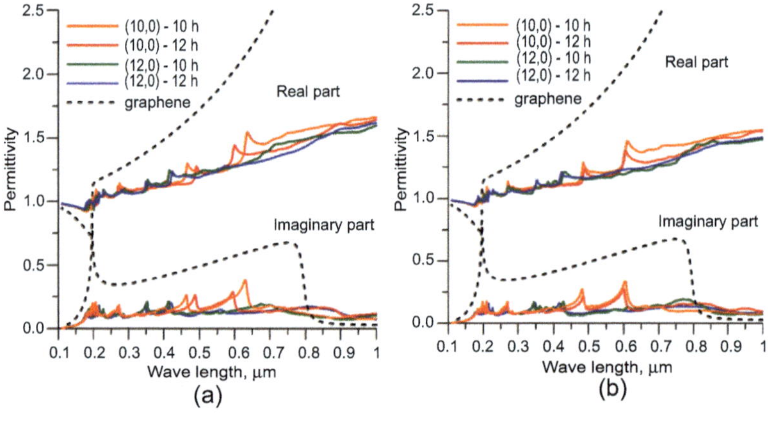

Figure 10.5 Graphs of complex dielectric permittivity for mono- (a) and bilayer (b) composite for the case of two tubes and different distances between them.

dependences for the same models as in Fig. 10.4. Graphs of the dielectric constant are shown in Fig. 10.5.

The calculated elements of the optical permittivity tensor were further used to estimate the reflectance and absorbance possibilities of the composite. To determine the reflection, transmission, and absorption parameters, Maxwell's equations for electric and magnetic fields in vacuum with a 2D CNT/graphene composite as the interface were solved. Figure 10.2 shows the variants of the incident wave vector plane configuration with respect to the atomic cell of the composite. In the general case, the wave vector **k** of a plane electromagnetic wave falling on a composite is located in the *xz* plane. The angle θ corresponds to the angle of incidence, and the vectors **E** and **H** correspond to the electric and magnetic field strengths, respectively. The electrodynamic medium is vacuum. The expressions for the reflection coefficient R and the transmission coefficient T in the case of a TE wave are written as follows:

$$R = \frac{-\sigma_{xx} Z_0}{2 + \sigma_{xx} Z_0}, \quad T = \frac{2}{2 + \sigma_{xx} Z_0} \tag{10.3}$$

For TH waves:

$$R = \frac{-\sigma_{yy} Z_0}{2 + \sigma_{yy} Z_0}, \quad T = \frac{2}{2 + \sigma_{yy} Z_0} \tag{10.4}$$

where Z_0 is the wave resistance in free space determined for p-polarized wave as $Z_0 = E_x/H_y = \eta\cos\theta$, for s-polarized wave as $Z_0 = -E_y/H_x = \eta/\cos\theta$; θ is the angle of incidence; $\eta = 120\pi\,\Omega$ is the input resistance of vacuum. With the expressions for the reflection and transmission coefficients, we can find the absorption coefficient by the following formula:

$$A_{p(s)} = 1 - R_{p(s)} - T_{p(s)} \tag{10.5}$$

Here, the index s/p indicates the type of wave. Very often optics uses the expression "s-wave" for TE waves and "p-wave" for TH waves.

Figure 10.6 presents the calculated values of the transparency coefficient for CNT/graphene films for normal incidence ($\theta = 0$) of electromagnetic waves. There are also two cases of polarization for different topological models of the composite. The spectra shown in Fig. 10.6 are divided into two groups. The solid curves mark the TE wave, and the dashed curves mark the TH waves. Differences between the families of curves for different types of waves show strong differences only in the range up to 2 μm; then they practically merge. It should be noted that the composite sharply differs from flat graphene. For graphene, distinction for different types of waves and the transparency spectra are qualitatively identical. The transparency of graphene, as is well known from experiment and

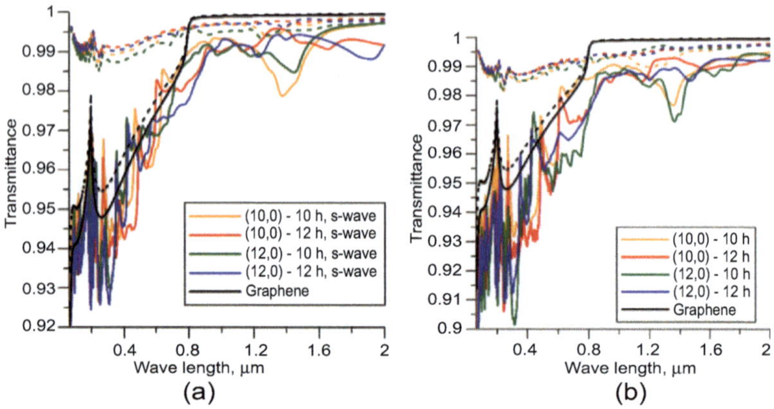

Figure 10.6 Spectra of the transparency coefficient for mono- (a) and bilayer (b) composite for the case of two tubes and different distances between them (solid curves mark TE wave, and dotted curves mark TH wave; here the expression "s-wave" corresponds to TE wave).

other theoretical works, has a minimum in the UV range, which provides 91–92% transparency at a wavelength of ~300 nm. At the same time, there is a sharp peak in this region. In the rest of the UV and visible region and in part of the IR region, graphene reflects 92–99%. The composite exhibits a large number of peaks in the spectrum. This is natural since the graph of the dynamic tensor conductivity elements is characterized by sharp peaks in the UV, visible, and IR regions. A large number of sharp peaks locate in the UV-visible regime, and they are clearly expressed for the TE wave (when the electric field vector is directed along the CNT). In the IR area, there are two laydowns for a composite with the step of 10 hexagons between the tubes. The highest peak is found in the UV region as for graphene. But the most important difference is that the transparency for the TH waves in the new composite is 99%, which is observed neither for individual tubes nor for graphene. This allows us to make a preliminary conclusion about the great prospects for using the composite as optical nanodevices. Indeed, a distinctive feature is the greater transparency of the composite for TH waves regardless of the number of layers. It is 98–100% for all wavelengths of the UV–visible–IR range.

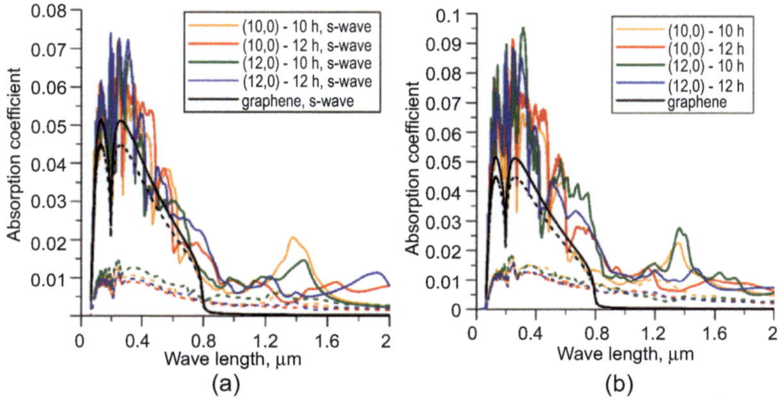

Figure 10.7 Spectra of the absorption coefficient for TE waves (solid line) and TH waves (dashed line): (a) monolayer composite and (b) bilayer composite.

The absorption spectra of the composite are shown in Fig. 10.7. For both types of waves, a sharp difference can be seen in the properties of the composite and graphene. The absorption of TH waves in the entire UV–visible–IR range is no more than 2%. Such a change in the absorptivity of the composite in comparison to ideal flat graphene is explained by its curvature in the direction perpendicular to the CNT axis due to covalent bonds with the tubes. For another type of waves in the UV range, the composite absorptivity is 1.5–2.0 times more but does not exceed 9%. It should be noted that there is no fundamental difference in the spectra for mono- and bilayer composites, i.e., an increase in the film thickness practically does not affect the optical characteristics of the composite. This can be seen from the data in Figs. 10.6 and 10.7.

Analysis of the calculated data of Figs. 10.6 and 10.7 shows that for all topological models of the composite, the characteristic peaks of high intensity are distinguished in the UV region at wavelengths of 150–350 nm (they are due to the influence of graphene) and in the optical range of 380–740 nm, regardless of the diameter of the tubes and the distance between them. There are no intense peaks in the IR region. The presence of high-intensity peaks for wavelengths of 150–250 nm is typical for graphene; so the presence of peaks in this region is inevitable in the UV region. However,

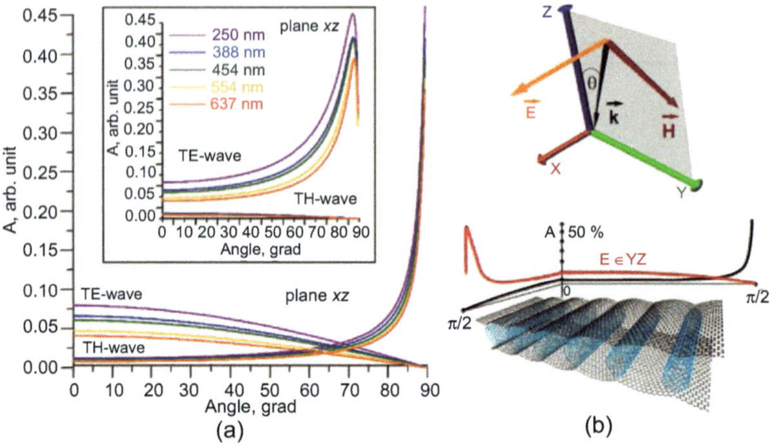

Figure 10.8 Diagrams of the absorption coefficient as a function of the incidence angle and the position of the wave vector (at different wavelengths): (a) graphs of the change in the coefficient A from the angle of the wave incidence; (b) relative position of the incidence plane and the fragment of the composite film.

the maximum absorption in graphene is less than in composite by almost 100%, that is, the composite is more promising for use in optical modulators than pure graphene.

For waves corresponding to the most intense peaks, we calculated the dependence of the absorption coefficient on the angle of incidence for cases when the wave vector lies in *xz* and *yz* planes. Figure 10.8 shows the change in the absorption coefficient for two kinds of wave polarization falling at different angles on a film of tubes (18,0) located between graphene sheets at the step of 13 hexagons. The diagrams were calculated for five cases of the incident wave: for wavelengths of 250, 388, 454, 524, and 637 nm. These values were selected in accordance with the calculated graphs similar to Fig. 10.7.

Four peaks in the optical range (at wavelengths of 388, 454, 524, and 637 nm) and a peak at a wavelength of 250 nm in the UV range, characteristic of all considered topological models of the composite, were selected. This choice was made by the application prospects of the investigated composite film as an operating element of an optical antenna or an optical modulator. The data in Fig. 10.8 show that for all wavelengths, the maximum absorption is observed in the TE wave at an incidence angle of 85–87° for the case of irradiation in the *xz* plane and for the TH wave at an incident angle of 85–90° for irradiation in the *yz* plane. At these angles of incidence, the absorption reaches a value of 45–50%.

Our results allow us to predict that few-layer hybrid composites based on graphene and CNTs are a promising material for transparent electronics in the UV–visible–IR range. For the middle part of the visible range (500 nm), red (700 nm) and violet (340 nm) border, we calculated the transparency coefficients of the TE and TH waves as a function of the composite film resistance. To calculate electric conductivity, the non-equilibrium Green's functions (NEGF) method was applied. We have used the standard Green function formalism to evaluate the conductance G of the system $G(E) = G_0 T(E)$ where $G_0 = e^2/h$ is the quantum of conductance, e is electron charge, and h is the Planck's constant. The transmission function $T(E)$ gives the probability of an electron to be transmitted at energy E from the source to the drain electrode. The electron transmission function $T(E)$ is calculated as follows:

$$T(E) = \mathrm{Tr}\left(g(E)\,\Gamma_{\mathrm{L}}(E)\,g^{*}(E)\,\Gamma_{\mathrm{LR}}(E)\right) \quad (10.6)$$

where

$$g(E) = \left[S_{\mathrm{M}}E^{+} - H_{\mathrm{M}} - \Sigma_{\mathrm{L}}(E) - \Sigma_{\mathrm{R}}(E)\right]^{-1} \quad (10.7)$$

$$\Gamma_{\{\mathrm{L,R}\}} = i\left[\Sigma_{\{\mathrm{L,R}\}} - \Sigma^{*}_{\{\mathrm{L,R}\}}\right] \quad (10.8)$$

Here Tr stands for the trace of the corresponding operator, H_{M} is the Hamiltonian matrix of the device, and S_{M} is the corresponding overlap matrix. $\Sigma_{\{\mathrm{L,R}\}}(E)$ are self-energy matrices for the left and right contacts. They are calculated according to the formula

$$\Sigma_{\{\mathrm{L,R}\}}(E) = \left(s_{\{\mathrm{L,R}\}}E^{+} - \tau_{\{\mathrm{L,R}\}}\right) g^{S}_{\{\mathrm{L,R}\}}(E) \times \left(s^{*}_{\{\mathrm{L,R}\}}E^{+} - \tau^{*}_{\{\mathrm{L,R}\}}\right), \quad (10.9)$$

where $\tau_{\{\mathrm{L,R}\}}$ are the coupling matrices between the active region and contacts, $s_{\{\mathrm{L,R}\}}$ are the corresponding overlap matrices, and $g^{S}_{\{\mathrm{L,R}\}}(E)$ are the surface Green functions of contacts [18].

The conductance of an object at low voltages can be determined by the transmission function [18]:

$$G = 2\frac{e^2}{h}\int T(E)\,F_{\mathrm{T}}(E)\,dE, \quad (10.10)$$

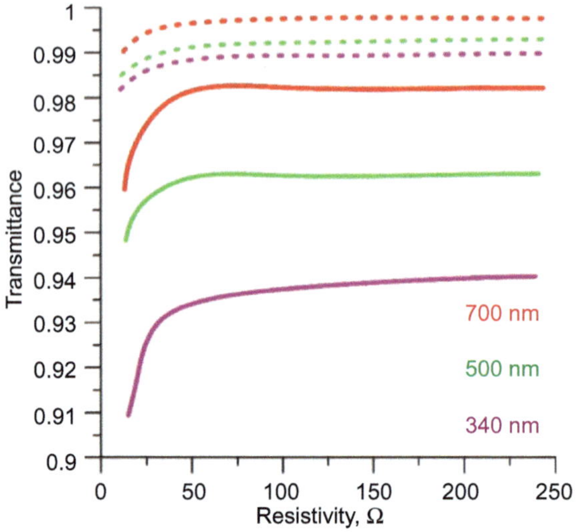

Figure 10.9 Dependence of the transparency coefficient on the resistance of the hybrid composite film (dashed lines correspond to the TH wave, and solid lines denote the TE wave).

$$F_{\mathrm{T}}(E) = \frac{1}{4K_{\mathrm{B}}T}\sec h^2\left(\frac{E-\mu}{K_{\mathrm{B}}T}\right), \tag{10.11}$$

where μ is the Fermi level of contacts, K_{B} is the Boltzmann constant, and T is the temperature. In our calculations, we used the self-consistent density functional tight-binding method (SCC DFTB). The usage of DFTB for the study of graphene-like structures with defects represents an extremely helpful compromise between accuracy and speed.

The calculated dependences are shown in Fig. 10.9. The general trend for both types of waves is a rapid increase in the transparency coefficient with an increase in resistance to 30 kΩ with further saturation. Saturation is achieved already at 110–120 kΩ. Further increase in film resistance does not affect its reflective possibility. The analysis of the calculated data shows that the composite films are most transparent for the wavelengths of the red region. However, the transparency of the films is judged by the wavelengths of the green colors corresponding to the middle of electromagnetic waves visible region. In this respect, we see that the transparency is

95–96% at the resistance of 0.025–0.25 kΩ. That is, the composite films are both highly conductive and transparent regardless of the incident wave type, i.e., of the wave vector incidence plane.

10.4 Conclusion

In general, it can be concluded that the optical properties of CNT–graphene 2D composite do not depend on the type of tube. The optical properties are almost similar for tubes with a metallic conductivity type and for tubes with semiconductor conductivity.

In summary, it can be concluded that a new composite material based on graphene and nanotubes is very promising for optoelectronic devices. In particular, the established regularities of change in the absorbance of the composite on the angle of incidence of the electromagnetic wave allow us to suggest the possibility of using a CNT–graphene film as a polarizer for electro-optical and magneto-optical thin film modulators. The advantages of such a polarizer will be a wide spectral range and low loss.

References

1. Mikhailov, S. A. (2016). Quantum theory of the third-order nonlinear electrodynamic effects of grapheme, *Phys. Rev. B*, **93**, 085403.
2. Cheng, J. L., Vermeulen, N., and Sipe, J. E. (2014). Third-order optical nonlinearity of grapheme, *New J. Phys.*, **16**, 053014.
3. Auditore, A., De Angelis, C., Locatelli, A., Boscolo, S., Midrio, M., Romagnoli, M., Capobianco, A. D., and Nalesso, G. (2013). Graphene-sustained nonlinear modes in dielectric waveguides, *Opt. Lett.*, **38**, pp. 631–633.
4. Hendry, E., Hale, P. J., Moger, J., Savchenko, A. K., and Mikhailov, S. A. (2010). Coherent nonlinear optical response of graphene, *Phys. Rev. Lett.*, **105**, 097401.
5. Maarouf, A. A., Kasry, A., Chandra, B., and Martyna, G. J. (2016). A graphene carbon nanotube hybrid material for photovoltaic applications, *Carbon*, **102**, pp. 74–80.

6. Gorkina, A. L., Tsapenko, A. P., Gilshteyn, E. P., Koltsova, T. S., Larionova, T. V., Talyzin, A., Anisimov, A. S., Anoshkin, I. V., Kauppinen, E. I., Tolochko, O. V., and Nasibulin, A. G. (2016). Transparent and conductive hybrid graphene/carbon nanotube films, *Carbon*, **100**, pp. 501–507.
7. Liu, Y., Wang, F., Wang, X., Wang, X., Flahaut, E., Liu, X., Li, Y., Wang, X., Xu, Y., Shi, Y., and Zhang, R. (2015). Planar carbon nanotube-graphene hybrid films for high-performance broadband photodetectors, *Nat. Commun.*, **6**, 8589.
8. Gan, X., Lv, R., Bai, J., Zhang, Z., Wei, J., Huang, Z. H., Zhu, H., Kang, F., and Terrones, M. (2015). Efficient photovoltaic conversion of graphene–carbon nanotube hybrid films grown from solid precursors, *2D Mater*, **2**, 034003.
9. Eckmann, A., Felten, A., Mishchenko, A., Britnell, L., Krupke, R., Novoselov, K. S., and Casiraghi C. (2012). Probing the nature of defects in graphene by Raman spectroscopy, *Nano Lett.*, **12**, pp. 3925–3930.
10. Kholmanov, I. N., Magnuson, C. W., Piner, R., Kim, J. Y., Aliev, A. E., Tan, C., Kim, T. Y., Zakhidov, A. A., Sberveglieri, G., Baughman, R. H., and Ruoff, R. S. (2015). Optical, electrical, and electromechanical properties of hybrid graphene/carbon nanotube films, *Adv. Mater.*, **27**, pp. 3053–3059.
11. Tristán-López, F., Morelos-Gómez, A., Vega-Díaz, S. M., García-Betancourt, M. L., Perea-López, N., Elías, A. L., Muramatsu, H., Cruz-Silva, R., Tsuruoka, S., Kim, Y. A., Hayahsi, T., Kaneko, K., Endo, M., and Terrones, M. (2013). Large area films of alternating graphene-carbon nanotube layers processed in water, *ACS Nano*, **7**, pp. 10788–10798.
12. Elstner, M., Porezag, D., Jungnickel, G., Elsner, J., Haugk, M., Frauenheim, Th., Suhai, S., and Seifert, G. (1998). Self-consistent-charge density-functional tight-binding method for simulations of complex materials properties, *Phys. Rev. B*, **58**, 7260.
13. Elstner, M. and Seifert, G. (2014). Density functional tight binding, *Phil. Trans. R. Soc. A*, **372**, 20120483.
14. Mitrofanov, V. V., Slepchenkov, M. M., Zhang, G., and Glukhova, O. E. (2017). Hybrid carbon nanotube-graphene monolayer films: Regularities of structure, electronic and optical properties, *Carbon*, **115**, pp. 803–810.
15. Marder, M. P. (2011). *Condensed Matter Physics*, 2nd ed. (Wiley-VCH: Berlin).
16. Le, H. A., Ho, S. T., Chien, N. D., and Do, V. N. (2014). Structure and spin dynamics of multiferroric $BiFeO_3$, *J. Phys. Condens. Mat.*, **26**, 405304.

17. Pedersen, T. G., Pedersen, K., and Kriestensen, T. B. (2001). Optical matrix elements in tight-binding calculations, *Phys. Rev. B*, **63**, 201101(R).
18. Datta, S. (2005). *Quantum Transport: Atom to Transistor*, 1st ed. (Cambridge University Press, Cambridge: New York).

Index